CAMBRIDGE LIBRARY COLLECTION

Books of enduring scholarly value

Darwin

Two hundred years after his birth and 150 years after the publication of 'On the Origin of Species', Charles Darwin and his theories are still the focus of worldwide attention. This series offers not only works by Darwin, but also the writings of his mentors in Cambridge and elsewhere, and a survey of the impassioned scientific, philosophical and theological debates sparked by his 'dangerous idea'.

Darwiniana

Darwiniana is a collection of critical essays on Charles Darwin's theory of evolution that were originally published in scientific journals by his friend and correspondent Asa Gray, Professor of Botany at Harvard. Gray was one of Darwin's strongest supporters in the American scientific community, and one of the few people with whom Darwin discussed his ideas in detail before publishing On the Origin of Species. Gray was also a Presbyterian and discussed questions of natural theology, design and teleology, including an earlier version of Chapter 3 of this book, with Darwin by letter over several years. Darwiniana (1876) was intended to provide a balanced assessment of Darwin's theory of evolution and to familiarise readers with the different aspects of Darwinism and its implications. The opening essays of the volume focus on the scientific and philosophical features of the theory, others analyse the reactions of Darwin's contemporaries and, most famously, argue for a reconciliation of religion and science in the light of Darwin's theory. The book will be a valuable resource for all who are interested in the American reception of Darwin's theory of evolution, and the scientific, philosophical and religious questions that it raised.

Cambridge University Press has long been a pioneer in the reissuing of out-of-print titles from its own backlist, producing digital reprints of books that are still sought after by scholars and students but could not be reprinted economically using traditional technology. The Cambridge Library Collection extends this activity to a wider range of books which are still of importance to researchers and professionals, either for the source material they contain, or as landmarks in the history of their academic discipline.

Drawing from the world-renowned collections in the Cambridge University Library, and guided by the advice of experts in each subject area, Cambridge University Press is using state-of-the-art scanning machines in its own Printing House to capture the content of each book selected for inclusion. The files are processed to give a consistently clear, crisp image, and the books finished to the high quality standard for which the Press is recognised around the world. The latest print-on-demand technology ensures that the books will remain available indefinitely, and that orders for single or multiple copies can quickly be supplied.

The Cambridge Library Collection will bring back to life books of enduring scholarly value across a wide range of disciplines in the humanities and social sciences and in science and technology.

Darwiniana

Essays and Reviews Pertaining to Darwinism

Asa Gray

CAMBRIDGE UNIVERSITY PRESS

Cambridge New York Melbourne Madrid Cape Town Singapore São Paolo Delhi

Published in the United States of America by Cambridge University Press, New York

www.cambridge.org
Information on this title: www.cambridge.org/9781108001960

This edition first published 1876
This digitally printed version 2009

ISBN 978-1-108-00196-0

DARWINIANA:

ESSAYS AND REVIEWS PERTAINING TO DARWINISM.

BY

ASA GRAY,

FISHER PROFESSOR OF NATURAL HISTORY (BOTANY) IN HARVARD UNIVERSITY.

NEW YORK:
D. APPLETON AND COMPANY,
549 & 551 BROADWAY.
1876.

PREFACE.

These papers are now collected at the request of friends and correspondents, who think that they may be useful; and two new essays are added. Most of the articles were written as occasion called for them within the past sixteen years, and contributed to various periodicals, with little thought of their forming a series, and none of ever bringing them together into a volume, although one of them (the third) was once reprinted in a pamphlet form. It is, therefore, inevitable that there should be considerable iteration in the argument, if not in the language. This could not be eliminated except by recasting the whole, which was neither practicable nor really desirable. It is better that they should record, as they do, the writer's freely-expressed thoughts upon the subject at the time; and to many readers there may be some advantage in going more than once, in different directions, over the same ground. If these essays were to be written now, some things might be differently expressed or qualified, but probably not so as to affect materially any important point. Accordingly, they are here reprinted unchanged, except by a few merely verbal alterations made in proof-reading, and the striking out of one or two superfluous or immaterial passages. A very few additional notes or references are appended.

To the last article but one a second part is now added, and the more elaborate Article XIII. is wholly new.

If it be objected that some of these pages are written in a lightness of vein not quite congruous with the gravity of the subject and the seriousness of its issues, the excuse must be that they were written with perfect freedom, most of them as anonymous contributions to popular journals, and that an argument may not be the less sound or an exposition less effective for being playful. Some of the essays, however, dealing with points of speculative scientific interest, may redress the balance, and be thought sufficiently heavy if not solid.

To the objection likely to be made, that they cover only a part of the ground, it can only be replied that they do not pretend to be systematic or complete. They are all essays relating in some way or other to the subject which has been, during these years, of paramount interest to naturalists, and not much less so to most thinking people. The first appeared between sixteen and seventeen years ago, immediately after the publication of Darwin's "Origin of Species by Means of Natural Selection," as a review of that volume, which, it was then foreseen, was to initiate a revolution in general scientific opinion. Long before our last article was written, it could be affirmed that the general doctrine of the derivation of species (to put it comprehensively) has prevailed over that of specific creation, at least to the extent of being the received and presumably in some sense true conception. Far from undertaking any general discussion of evolution, several even of Mr. Darwin's writings have not been noticed, and topics which have been much discussed elsewhere are not here adverted to. This applies especially to what may be called deductive evolution—a subject which lay beyond the writer's immediate scope, and to which neither the bent of his mind nor the line of his studies has fitted him to do justice. If these papers are useful at all, it will be as showing how these new views of our day are regarded by a practical naturalist, versed in one department only (viz., Botany), most interested in their bearings upon its special problems, one accustomed to direct and close dealing with the facts in hand, and disposed to rise from them only to the consideration of those general questions upon which they throw or from which they receive illustration.

Then as to the natural theological questions which (owing to circumstances needless now to be recalled or explained) are here throughout brought into what most naturalists, and some other readers, may deem undue prominence, there are many who may be interested to know how these increasingly prevalent views and their tendencies are regarded by one who is scientifically, and in his own fashion, a Darwinian, philosophically a convinced theist, and religiously an acceptor of the "creed commonly called the Nicene," as the exponent of the Christian faith.

"Truth emerges sooner from error than from confusion," says Bacon; and clearer views than commonly prevail upon the points at issue regarding "religion and science" are still sufficiently needed to justify these endeavors.

BOTANIC GARDEN, CAMBRIDGE, MASS., *June*, 1876.

CONTENTS.

*** This Table of Contents, and the copious Index to the volume, were obligingly prepared by the Rev. G. F. WRIGHT, of Andover.

ARTICLE I.

THE ORIGIN OF SPECIES BY MEANS OF NATURAL SELECTION.

ARTICLE II.

DESIGN *versus* NECESSITY—A DISCUSSION.

ARTICLE III.

NATURAL SELECTION NOT INCONSISTENT WITH NATURAL THEOLOGY.

ARTICLE IV.

SPECIES AS TO VARIATION, GEOGRAPHICAL DISTRIBUTION, AND SUCCESSION.

ARTICLE V.

SEQUOIA AND ITS HISTORY: THE RELATIONS OF NORTH AMERICAN TO NORTH-EASTERN ASIAN AND TO TERTIARY VEGETATION.

ARTICLE VI.

THE ATTITUDE OF WORKING NATURALISTS TOWARD DARWINISM.

ARTICLE VII.

EVOLUTION AND THEOLOGY.

ARTICLE VIII.

"WHAT IS DARWINISM?"

ARTICLE XII.

DURATION AND ORIGINATION OF RACE AND SPECIES.

ARTICLE XIII.

EVOLUTIONARY TELEOLOGY.

DARWINIANA.

I.

THE ORIGIN OF SPECIES BY MEANS OF NATURAL SELECTION.[1]

(American Journal of Science and Arts, *March*, 1860.)

This book is already exciting much attention. Two American editions are announced, through which it will become familiar to many of our readers, before these pages are issued. An abstract of the argument —for "the whole volume is one long argument," as the author states—is unnecessary in such a case; and it would be difficult to give by detached extracts. For the volume itself is an abstract, a prodromus of a detailed work upon which the author has been laboring for twenty years, and which "will take two or three more years to complete." It is exceedingly compact; and although useful summaries are appended to the several chapters, and a general recapitulation con-

[1] "On the Origin of Species by Means of Natural Selection, or the Preservation of Favored Races in the Struggle for Life," by Charles Darwin, M. A., Fellow of the Royal, Geological, Linnæan, etc., Societies, Author of "Journal of Researches during H. M. S. Beagle's Voyage round the World." London: John Murray. 1859. 502 pp., post 8vo.

tains the essence of the whole, yet much of the aroma escapes in the treble distillation, or is so concentrated that the flavor is lost to the general or even to the scientific reader. The volume itself—the proof-spirit—is just condensed enough for its purpose. It will be far more widely read, and perhaps will make deeper impression, than the elaborate work might have done, with all its full details of the facts upon which the author's sweeping conclusions have been grounded. At least it is a more readable book: but all the facts that can be mustered in favor of the theory are still likely to be needed.

Who, upon a single perusal, shall pass judgment upon a work like this, to which twenty of the best years of the life of a most able naturalist have been devoted? And who among those naturalists who hold a position that entitles them to pronounce summarily upon the subject, can be expected to divest himself for the nonce of the influence of received and favorite systems? In fact, the controversy now opened is not likely to be settled in an off-hand way, nor is it desirable that it should be. A spirited conflict among opinions of every grade must ensue, which—to borrow an illustration from the doctrine of the book before us—may be likened to the conflict in Nature among races in the struggle for life, which Mr. Darwin describes; through which the views most favored by facts will be developed and tested by "Natural Selection," the weaker ones be destroyed in the process, and the strongest in the long-run alone survive.

The duty of reviewing this volume in the *American Journal of Science* would naturally devolve upon the principal editor, whose wide observation and profound knowledge of various departments of natural history, as well as of geology, particularly qualify him for the task. But he has been obliged to lay aside his pen, and to seek in distant lands the entire repose from scientific labor so essential to the restoration of his health—a consummation devoutly to be wished, and confidently to be expected. Interested as Mr. Dana would be in this volume, he could not be expected to accept its doctrine. Views so idealistic as those upon which his "Thoughts upon Species"[1] are grounded, will not harmonize readily with a doctrine so thoroughly naturalistic as that of Mr. Darwin. Though it is just possible that one who regards the kinds of elementary matter, such as oxygen and hydrogen, and the definite compounds of these elementary matters, and their compounds again, in the mineral kingdom, as constituting species, in the same sense, fundamentally, as that of animal and vegetable species, might admit an evolution of one species from another in the latter as well as the former case.

Between the doctrines of this volume and those of the other great naturalist whose name adorns the title-page of this journal [Mr. Agassiz], the widest divergence appears. It is interesting to contrast the two, and, indeed, is necessary to our purpose; for this contrast brings out most prominently, and sets in strongest light and shade, the main features of the theory of the origination of species by means of Natural Selection.

The ordinary and generally-received view assumes the independent, specific creation of each kind of plant

[1] Article in this Journal, vol. xxiv., p. 305.

and animal in a primitive stock, which reproduces its like from generation to generation, and so continues the species.[1] Taking the idea of species from this perennial succession of essentially similar individuals, the chain is logically traceable back to a local origin in a single stock, a single pair, or a single individual, from which all the individuals composing the species have proceeded by natural generation. Although the similarity of progeny to parent is fundamental in the conception of species, yet the likeness is by no means absolute; all species vary more or less, and some vary remarkably—partly from the influence of altered circumstances, and partly (and more really) from unknown constitutional causes which altered conditions favor rather than originate. But these variations are supposed to be mere oscillations from a normal state, and in Nature to be limited if not transitory; so that the primordial differences between species and species at their beginning have not been effaced, nor largely obscured, by blending through variation. Consequently, whenever two reputed species are found to blend in Nature through a series of intermediate forms, community of origin is inferred, and all the forms, however diverse, are held to belong to one species. Moreover, since bisexuality is the rule in Nature (which is practically carried out, in the long-run, far more generally than has been suspected), and the heritable qualities of two distinct individuals are mingled in the offspring, it is supposed that the general

[1] "Species tot sunt, quot diversas formas ab initio produxit Infinitum Ens; quæ formæ, secundum generationis inditas leges, produxere plures, at sibi semper similes."—*Linn. Phil. Bot.*, 99, 157.

sterility of hybrid progeny interposes an effectual barrier against the blending of the original species by crossing.

From this generally-accepted view the well-known theory of Agassiz and the recent one of Darwin diverge in exactly opposite directions.

That of Agassiz differs fundamentally from the ordinary view only in this, that it discards the idea of a common descent as the real bond of union among the individuals of a species, and also the idea of a local origin—supposing, instead, that each species originated simultaneously, generally speaking, over the whole geographical area it now occupies or has occupied, and in perhaps as many individuals as it numbered at any subsequent period.

Mr. Darwin, on the other hand, holds the orthodox view of the descent of all the individuals of a species not only from a local birthplace, but from a single ancestor or pair; and that each species has extended and established itself, through natural agencies, wherever it could; so that the actual geographical distribution of any species is by no means a primordial arrangement, but a natural result. He goes farther, and this volume is a protracted argument intended to prove that the species we recognize have not been independently created, as such, but have descended, like varieties, from other species. Varieties, on this view, are incipient or possible species: species are varieties of a larger growth and a wider and earlier divergence from the parent stock; the difference is one of degree, not of kind.

The ordinary view—rendering unto Cæsar the

things that are Cæsar's—looks to natural agencies for the actual distribution and perpetuation of species, to a supernatural for their origin.

The theory of Agassiz regards the origin of species and their present general distribution over the world as equally primordial, equally supernatural; that of Darwin, as equally derivative, equally natural.

The theory of Agassiz, referring as it does the phenomena both of origin and distribution directly to the Divine will—thus removing the latter with the former out of the domain of inductive science (in which efficient cause is not the first, but the last word) —may be said to be theistic to excess. The contrasted theory is not open to this objection. Studying the facts and phenomena in reference to proximate causes, and endeavoring to trace back the series of cause and effect as far as possible, Darwin's aim and processes are strictly scientific, and his endeavor, whether successful or futile, must be regarded as a legitimate attempt to extend the domain of natural or physical science. For, though it well may be that "organic forms have no physical or secondary cause," yet this can be proved only indirectly, by the failure of every attempt to refer the phenomena in question to causal laws. But, however originated, and whatever be thought of Mr. Darwin's arduous undertaking in this respect, it is certain that plants and animals are subject from their birth to physical influences, to which they have to accommodate themselves as they can. How literally they are "born to trouble," and how incessant and severe the struggle for life generally is, the present volume graphically describes. Few will deny that such influences must have gravely affected the range and the association of individuals and species on the earth's surface. Mr. Darwin thinks that, acting upon an inherent predisposition to vary, they have sufficed even to modify the species themselves and produce the present diversity. Mr. Agassiz believes that they have not even affected the geographical range and the actual association of species, still less their forms; but that every adaptation of species to climate, and of species to species, is as aboriginal, and therefore as inexplicable, as are the organic forms themselves.

Who shall decide between such extreme views so ably maintained on either hand, and say how much of truth there may be in each? The present reviewer has not the presumption to undertake such a task. Having no prepossession in favor of naturalistic theories, but struck with the eminent ability of Mr. Darwin's work, and charmed with its fairness, our humbler duty will be performed if, laying aside prejudice as much as we can, we shall succeed in giving a fair account of its method and argument, offering by the way a few suggestions, such as might occur to any naturalist of an inquiring mind. An editorial character for this article must in justice be disclaimed. The plural pronoun is employed not to give editorial weight, but to avoid even the appearance of egotism, and also the circumlocution which attends a rigorous adherence to the impersonal style.

We have contrasted these two extremely divergent theories, in their broad statements. It must not be inferred that they have no points nor ultimate results in common.

In the first place, they practically agree in upsetting, each in its own way, the generally-received definition of species, and in sweeping away the ground of their objective existence in Nature. The orthodox conception of species is that of lineal descent: all the descendants of a common parent, and no other, constitute a species; they have a certain identity because of their descent, by which they are supposed to be recognizable. So naturalists had a distinct idea of what they meant by the term species, and a practical rule, which was hardly the less useful because difficult to apply in many cases, and because its application was indirect: that is, the community of origin had to be inferred from the likeness; such degree of similarity, and such only, being held to be conspecific as could be shown or reasonably inferred to be compatible with a common origin. And the usual concurrence of the whole body of naturalists (having the same data before them) as to what forms are species attests the value of the rule, and also indicates some real foundation for it in Nature. But if species were created in numberless individuals over broad spaces of territory, these individuals are connected only in idea, and species differ from varieties on the one hand, and from genera, tribes, etc., on the other, only in degree; and no obvious natural reason remains for fixing upon this or that degree as specific, at least no natural standard, by which the opinions of different naturalists may be correlated. Species upon this view are enduring, but subjective and ideal. Any three or more of the human races, for example, are species or not species, according to the bent of the naturalist's mind. Darwin's theory brings us the other way to the same result. In his view, not only all the individuals of a species are descendants of a common parent, but of all the related species also. Affinity, relationship, all the terms which naturalists use figuratively to express an underived, unexplained resemblance among species, have a literal meaning upon Darwin's system, which they little suspected, namely, that of inheritance. Varieties are the latest offshoots of the genealogical tree in "an unlineal" order; species, those of an earlier date, but of no definite distinction; genera, more ancient species, and so on. The human races, upon this view, likewise may or may not be species according to the notions of each naturalist as to what differences are specific; but, if not species already, those races that last long enough are sure to become so. It is only a question of time.

How well the simile of a genealogical tree illustrates the main ideas of Darwin's theory the following extract from the summary of the fourth chapter shows:

"It is a truly wonderful fact—the wonder of which we are apt to overlook from familiarity—that all animals and all plants throughout all time and space should be related to each other in group subordinate to group, in the manner which we everywhere behold—namely, varieties of the same species most closely related together, species of the same genus less closely and unequally related together, forming sections and sub-genera, species of distinct genera much less closely related, and genera related in different degrees, forming sub-families, families, orders, sub-classes, and classes. The several subordinate groups in any class cannot be ranked in a single file, but seem rather to be clustered round points, and these round other points, and so on in almost endless cycles. On the view that each species has been independently created, I can see no explanation of this great fact in the classification of all organic beings; but, to the best of my judgment, it is explained through inheritance and the complex action of natural selection, entailing extinction and divergence of character, as we have seen illustrated in the diagram.

"The affinities of all the beings of the same class have sometimes been represented by a great tree. I believe this simile largely speaks the truth. The green and budding twigs may represent existing species; and those produced during each former year may represent the long succession of extinct species. At each period of growth all the growing twigs have tried to branch out on all sides, and overtop and kill the surrounding twigs and branches, in the same manner as species and groups of species have tried to overmaster other species in the great battle for life. The limbs divided into great branches, and these into lesser and lesser branches, were themselves once, when the tree was small, budding twigs; and this connection of the former and present buds by ramifying branches may well represent the classification of all extinct and living species in groups subordinate to groups. Of the many twigs which flourished when the tree was a mere bush, only two or three, now grown into great branches, yet survive and bear all the other branches; so with the species which lived during long-past geological periods, very few now have living and modified descendants. From the first growth of the tree, many a limb and branch has decayed and dropped off; and these lost branches of various sizes may represent those whole orders, families, and genera, which have now no living representatives, and which are known to us only from having been found in a fossil state. As we here and there see a thin, straggling branch springing from a fork low down in a tree, and which by some chance has been favored and is still alive on its summit, so we occasionally see an animal like the Ornithorhynchus or Lepidosiren, which in some small degree connects by its affinities two large branches of life, and which has apparently been saved from fatal competition by having inhabited a protected station. As buds give rise by growth to fresh buds, and these, if vigorous, branch out and overtop on all sides many a feebler branch, so by generation I believe it has been with the great Tree of Life, which fills with its dead and broken branches the crust of the earth, and covers the surface with its ever-branching and beautiful ramifications."

It may also be noted that there is a significant correspondence between the rival theories as to the main facts employed. Apparently every capital fact in the one view is a capital fact in the other. The difference is in the interpretation. To run the parallel ready made to our hands:[1]

"The simultaneous existence of the most diversified types under identical circumstances, ... the repetition of similar types under the most diversified circumstances, ... the unity of plan in otherwise highly-diversified types of animals, ... the correspondence, now generally known as special homologies, in the details of structure otherwise entirely disconnected, down to the most minute peculiarities, ... the various degrees and different kinds of relationship among animals which (apparently) can have no genealogical connection, ... the simultaneous existence in the earliest geological periods, ... of representatives of all the great types of the animal kingdom, ... the gradation based upon complications of structure which may be traced among animals built upon the same plan; the distribution of some types over the most extensive range of surface of the globe, while others are limited to particular geographical areas, ... the identity of structures of these types, notwithstanding their wide geographical distribution, ... the community of structure in certain respects of animals otherwise entirely different, but living within the same geographical area, ... the connection by series of special structures observed in animals widely scattered over the surface of the globe, ... the definite relations in which animals stand to the surrounding world, ... the relations in which individuals of the same

[1] Agassiz, "Essay on Classification; Contributions to Natural History," p. 132, *et seq.*

species stand to one another, ... the limitation of the range of changes which animals undergo during their growth, ... the return to a definite norm of animals which multiply in various ways, ... the order of succession of the different types of animals and plants characteristic of the different geological epochs, ... the localization of some types of animals upon the same points of the surface of the globe during several successive geological periods, ... the parallelism between the order of succession of animals and plants in geological times, and the gradation among their living representatives, ... the parallelism between the order of succession of animals in geological times and the changes their living representatives undergo during their embryological growth,[1] . . . . *the combination in many extinct types of characters which in later ages appear disconnected in different types*, ... the parallelism between the gradation among animals and the changes they undergo during their growth, ... the relations existing between these different series and the geographical distribution of animals, ... the connection of all the known features of Nature into one system—"

In a word, the whole relations of animals, etc., to surrounding Nature and to each other, are regarded under the one view as ultimate facts, or in their ultimate aspect, and interpreted theologically; under the other as complex facts, to be analyzed and interpreted

[1] As to this, Darwin remarks that he can only hope to see the law hereafter proved true (p. 449); and p. 338: "Agassiz insists that ancient animals resemble to a certain extent the embryos of recent animals of the same classes; or that the geological succession of extinct forms is in some degree parallel to the embryological development of recent forms. I must follow Pictet and Huxley in thinking that the truth of this doctrine is very far from proved. Yet I fully expect to see it hereafter confirmed, at least in regard to subordinate groups, which have branched off from each other within comparatively recent times. For this doctrine of Agassiz accords well with the theory of natural selection."

scientifically. The one naturalist, perhaps too largely assuming the scientifically unexplained to be inexplicable, views the phenomena only in their supposed relation to the Divine mind. The other, naturally expecting many of these phenomena to be resolvable under investigation, views them in their relations to one another, and endeavors to explain them as far as he can (and perhaps farther) through natural causes.

But does the one really exclude the other? Does the investigation of physical causes stand opposed to the theological view and the study of the harmonies between mind and Nature? More than this, is it not most presumable that an intellectual conception realized in Nature would be realized through natural agencies? Mr. Agassiz answers these questions affirmatively when he declares that "the task of science is to investigate what has been done, to inquire if possible *how it has been done*, rather than to ask what is possible for the Deity, since *we can know that only by what actually exists;*" and also when he extends the argument for the intervention in Nature of a creative mind to its legitimate application in the inorganic world; which, he remarks, "considered in the same light, would not fail also to exhibit unexpected evidence of thought, in the character of the laws regulating the chemical combinations, the action of physical forces, etc., etc."[1] Mr. Agassiz, however, pronounces that "the connection between the facts is *only intellectual*"—an opinion which the analogy of the inor-

[1] *Op. cit.*, p. 131.—One or two Bridgewater Treatises, and most modern works upon natural theology, should have rendered the evidences of thought in inorganic Nature not "unexpected."

ganic world, just referred to, does not confirm, for there a material connection between the facts is justly held to be consistent with an intellectual—and which the most analogous cases we can think of in the organic world do not favor; for there is a material connection between the grub, the pupa, and the butterfly, between the tadpole and the frog, or, still better, between those distinct animals which succeed each other in alternate and very dissimilar generations. So that mere analogy might rather suggest a natural connection than the contrary; and the contrary cannot be demonstrated until the possibilities of Nature under the Deity are fathomed.

But, the intellectual connection being undoubted, Mr. Agassiz properly refers the whole to "the agency of Intellect as its first cause." In doing so, however, he is not supposed to be offering a scientific explanation of the phenomena. Evidently he is considering only the ultimate *why*, not the proximate why or *how*.

Now the latter is just what Mr. Darwin is considering. He conceives of a physical connection between allied species; but we suppose he does not deny their intellectual connection, as related to a supreme intelligence. Certainly we see no reason why he should, and many reasons why he should not. Indeed, as we contemplate the actual direction of investigation and speculation in the physical and natural sciences, we dimly apprehend a probable synthesis of these divergent theories, and in it the ground for a strong stand against mere naturalism. Even if the doctrine of the origin of species through natural selection should prevail in our day, we shall not despair; being confident that the genius of an Agassiz will be found equal to the work of constructing, upon the mental and material foundations combined, a theory of Nature as theistic and as scientific as that which he has so eloquently expounded.

To conceive the possibility of "the descent of species from species by insensibly fine gradations" during a long course of time, and to demonstrate its compatibility with a strictly theistic view of the universe, is one thing; to substantiate the theory itself or show its likelihood is quite another thing. This brings us to consider what Darwin's theory actually is, and how he supports it.

That the existing kinds of animals and plants, or many of them, may be derived from other and earlier kinds, in the lapse of time, is by no means a novel proposition. Not to speak of ancient speculations of the sort, it is the well-known Lamarckian theory. The first difficulty which such theories meet with is that in the present age, with all its own and its inherited prejudgments, the whole burden of proof is naturally, and indeed properly, laid upon the shoulders of the propounders; and thus far the burden has been more than they could bear. From the very nature of the case, substantive proof of specific creation is not attainable; but that of derivation or transmutation of species may be. He who affirms the latter view is bound to do one or both of two things: 1. Either to assign real and adequate causes, the natural or necessary result of which must be to produce the present diversity of species and their actual relations; or, 2. To show the general conformity of the whole body of facts to such assumption, and also to adduce instances explicable by it and inexplicable by the received view, so perhaps winning our assent to the doctrine, through its competency to harmonize all the facts, even though the cause of the assumed variation remain as occult as that of the transformation of tadpoles into frogs, or that of *Coryne* into *Sarzia*.

The first line of proof, successfully carried out, would establish derivation as a true physical theory; the second, as a sufficient hypothesis.

Lamarck mainly undertook the first line, in a theory which has been so assailed by ridicule that it rarely receives the credit for ability to which in its day it was entitled. But he assigned partly unreal, partly insufficient causes; and the attempt to account for a progressive change in species through the direct influence of physical agencies, and through the appetencies and habits of animals reacting upon their structure, thus causing the production and the successive modification of organs, is a conceded and total failure. The shadowy author of the "Vestiges of the Natural History of Creation" can hardly be said to have undertaken either line, in a scientific way. He would explain the whole progressive evolution of Nature by virtue of an inherent tendency to development, thus giving us an idea or a word in place of a natural cause, a restatement of the proposition instead of an explanation. Mr. Darwin attempts both lines of proof, and in a strictly scientific spirit; but the stress falls mainly upon the first, for, as he does assign real causes, he is bound to prove their adequacy.

It should be kept in mind that, while all direct

proof of independent origination is attainable from the nature of the case, the overthrow of particular schemes of derivation has not established the opposite proposition. The futility of each hypothesis thus far proposed to account for derivation may be made apparent, or unanswerable objections may be urged against it; and each victory of the kind may render derivation more improbable, and therefore specific creation more probable, without settling the question either way. New facts, or new arguments and a new mode of viewing the question, may some day change the whole aspect of the case. It is with the latter that Mr. Darwin now reopens the discussion.

Having conceived the idea that varieties are incipient species, he is led to study variation in the field where it shows itself most strikingly, and affords the greatest facilities to investigation. Thoughtful naturalists have had increasing grounds to suspect that a reëxamination of the question of species in zoölogy and botany, commencing with those races which man knows most about, viz., the domesticated and cultivated races, would be likely somewhat to modify the received idea of the entire fixity of species. This field, rich with various but unsystematized stores of knowledge accumulated by cultivators and breeders, has been generally neglected by naturalists, because these races are not in a state of nature; whereas they deserve particular attention on this very account, as experiments, or the materials for experiments, ready to our hand. In domestication we vary some of the natural conditions of a species, and thus learn experimentally what changes are within the reach of vary ing conditions in Nature. We separate and protect a favorite race against its foes or its competitors, and thus learn what it might become if Nature ever afforded it equal opportunities. Even when, to subserve human uses, we modify a domesticated race to the detriment of its native vigor, or to the extent of practical monstrosity, although we secure forms which would not be originated and could not be perpetuated in free Nature, yet we attain wider and juster views of the possible degree of variation. We perceive that some species are more variable than others, but that no species subjected to the experiment persistently refuses to vary; and that, when it has once begun to vary, its varieties are not the less but the more subject to variation. "No case is on record of a variable being ceasing to be variable under cultivation." It is fair to conclude, from the observation of plants and animals in a wild as well as domesticated state, that the tendency to vary is general, and even universal. Mr. Darwin does "not believe that variability is an inherent and necessary contingency, under all circumstances, with all organic beings, as some authors have thought." No one supposes variation could occur under all circumstances; but the facts on the whole imply a universal tendency, ready to be manifested under favorable circumstances. In reply to the assumption that man has chosen for domestication animals and plants having an extraordinary inherent tendency to vary, and likewise to withstand diverse climates, it is asked:

"How could a savage possibly know, when he first tamed an animal, whether it would vary in succeeding generations, and whether it would endure other climates? Has the little variability of the ass or Guinea-fowl, or the small power of endurance of warmth by the reindeer, or of cold by the common camel, prevented their domestication? I cannot doubt that if other animals and plants, equal in number to our domesticated productions, and belonging to equally diverse classes and countries, were taken from a state of nature, and could be made to breed for an equal number of generations under domestication, they would vary on an average as largely as the parent species of our existing domesticated productions have varied."

As to amount of variation, there is the common remark of naturalists that the varieties of domesticated plants or animals often differ more widely than do the individuals of distinct species in a wild state: and even in Nature the individuals of some species are known to vary to a degree sensibly wider than that which separates related species. In his instructive section on the breeds of the domestic pigeon, our author remarks that "at least a score of pigeons might be chosen which if shown to an ornithologist, and he were told that they were wild birds, would certainly be ranked by him as well-defined species. Moreover, I do not believe that any ornithologist would place the English carrier, the short-faced tumbler, the runt, the barb, pouter, and fantail, in the same genus; more especially as in each of these breeds several truly-inherited sub-breeds, or species, as he might have called them, could be shown him." That this is not a case like that of dogs, in which probably the blood of more than one species is mingled, Mr. Darwin proceeds to show, adducing cogent reasons for the common opinion that all have descended from the wild rock-pigeon. Then follow some suggestive remarks:

"I have discussed the probable origin of domestic pigeons at some, yet quite insufficient, length; because when I first kept pigeons and watched the several kinds, knowing well how true they bred, I felt fully as much difficulty in believing that they could ever have descended from a common parent as any naturalist could in coming to a similar conclusion in regard to many species of finches, or other large groups of birds, in Nature. One circumstance has struck me much; namely, that all the breeders of the various domestic animals and the cultivators of plants, with whom I have ever conversed, or whose treatises I have read, are firmly convinced that the several breeds to which each has attended are descended from so many aboriginally distinct species. Ask, as I have asked, a celebrated raiser of Hereford cattle, whether his cattle might not have descended from long-horns, and he will laugh you to scorn. I have never met a pigeon, or poultry, or duck, or rabbit fancier, who was not fully convinced that each main breed was descended from a distinct species. Van Mons, in his treatise on pears and apples, shows how utterly he disbelieves that the several sorts, for instance a Ribston-pippin or Codlin-apple, could ever have proceeded from the seeds of the same tree. Innumerable other examples could be given. The explanation, I think, is simple: from long-continued study they are strongly impressed with the differences between the several races; and though they well know that each race varies slightly, for they win their prizes by selecting such slight differences, yet they ignore all general arguments, and refuse to sum up in their minds slight differences accumulated during many successive generations. May not those naturalists who, knowing far less of the laws of inheritance than does the breeder, and knowing no more than he does of the intermediate links in the long lines of descent, yet admit that many of our domestic races have descended from the same parents—may they not learn a lesson of caution, when they deride the idea of species in a state of nature being lineal descendants of other species?"

The actual causes of variation are unknown. Mr. Darwin favors the opinion of the late Mr. Knight, the great philosopher of horticulture, that variability under domestication is somehow connected with excess of food. He regards the unknown cause as acting chiefly upon the reproductive system of the parents, which system, judging from the effect of confinement or cultivation upon its functions, he concludes to be more susceptible than any other to the action of changed conditions of life. The tendency to vary certainly appears to be much stronger under domestication than in free Nature. But we are not sure that the greater variableness of cultivated races is not mainly owing to the far greater opportunities for manifestation and accumulation—a view seemingly all the more favorable to Mr. Darwin's theory. The actual amount of certain changes, such as size or abundance of fruit, size of udder, stands of course in obvious relation to supply of food.

Really, we no more know the reason why the progeny occasionally deviates from the parent than we do why it usually resembles it. Though the laws and conditions governing variation are known to a certain extent, those governing inheritance are apparently inscrutable. "Perhaps," Darwin remarks, "the correct way of viewing the whole subject would be, to look at the inheritance of every character whatever as the rule, and non-inheritance as the anomaly." This, from general and obvious considerations, we have long been accustomed to do. Now, as exceptional instances are expected to be capable of explanation, while ultimate laws are not, it is quite possible that variation may be accounted for, while the great primary law of inheritance remains a mysterious fact.

The common proposition is, that *species reproduce their like;* this is a sort of general inference, only a degree closer to fact than the statement that genera reproduce their like. The true proposition, the fact incapable of further analysis, is, that *individuals reproduce their like*—that characteristics are inheritable. So varieties, or deviations, once originated, are perpetuable, like species. Not so likely to be perpetuated, at the outset; for the new form tends to resemble a grandparent and a long line of similar ancestors, as well as to resemble its immediate progenitors. Two forces which coincide in the ordinary case, where the offspring resembles its parent, act in different directions when it does not and it is uncertain which will prevail. If the remoter but very potent ancestral influence predominates, the variation disappears with the life of the individual. If that of the immediate parent—feebler no doubt, but closer—the variety survives in the offspring; whose progeny now has a redoubled tendency to produce its own like; whose progeny again is almost sure to produce its like, since it is much the same whether it takes after its mother or its grandmother.

In this way races arise, which under favorable conditions may be as hereditary as species. In following these indications, watching opportunities, and breeding only from those individuals which vary most in a desirable direction, man leads the course of variation as he leads a streamlet—apparently at will, but never against the force of gravitation—to a long distance from its source, and makes it more subservient to his use or fancy. He unconsciously strengthens those variations which he prizes when he plants the seed of a favorite fruit, preserves a favorite domestic animal, drowns the uglier kittens of a litter, and allows only the handsomest or the best mousers to propagate. Still more, by methodical selection, in recent times almost marvelous results have been produced in new breeds of cattle, sheep, and poultry, and new varieties of fruit of greater and greater size or excellence.

It is said that all domestic varieties, if left to run wild, would revert to their aboriginal stocks. Probably they would wherever various races of one species were left to commingle. At least the abnormal or exaggerated characteristics induced by high feeding, or high cultivation and prolonged close breeding, would promptly disappear; and the surviving stock would soon blend into a homogeneous result (in a way presently explained), which would naturally be taken for the original form; but we could seldom know if it were so. It is by no means certain that the result would be the same if the races ran wild each in a separate region. Dr. Hooker doubts if there is a true reversion in the case of plants. Mr. Darwin's observations rather favor it in the animal kingdom. With mingled races reversion seems well made out in the case of pigeons. The common opinion upon this subject therefore probably has some foundation. But even if we regard varieties as oscillations around a primitive centre or type, still it appears from the readiness with which such varieties originate that a certain amount of disturbance would carry them beyond the influence of the primordial attraction, where they may become new centres of variation.

Some suppose that races cannot be perpetuated indefinitely even by keeping up the conditions under which they were fixed; but the high antiquity of several, and the actual fixity of many of them, negative this assumption. "To assert that we could not breed our cart and race horses, long and short horned cattle, and poultry of various breeds, for almost an infinite number of generations, would be opposed to all experience."

Why varieties develop so readily and deviate so widely under domestication, while they are apparently so rare or so transient in free Nature, may easily be shown. In Nature, even with hermaphrodite plants, there is a vast amount of cross-fertilization among various individuals of the same species. The inevitable result of this (as was long ago explained in this Journal[1]) is to repress variation, to keep the mass of a species comparatively homogeneous over any area in which it abounds in individuals. Starting from a suggestion of the late Mr. Knight, now so familiar, that close interbreeding diminishes vigor and fertility;[2] and perceiving that bisexuality is ever aimed at in Nature—being attained physiologically in numerous cases where it is not structurally—Mr. Darwin has worked out the subject in detail, and shown how general is the concurrence, either habitual or occasional, of two hermaphrodite individuals in the reproduction of their kind; and has drawn the philosophical infer-

[1] Volume xvii. (2), 1854, p. 13.

[2] We suspect that this is not an ultimate fact, but a natural consequence of inheritance—the inheritance of disease or of tendency to disease, which close interbreeding perpetuates and accumulates, but wide breeding may neutralize or eliminate.

ence that probably no organic being self-fertilizes indefinitely; but that a cross with another individual is occasionally—perhaps at very long intervals—indispensable. We refer the reader to the section on the intercrossing of individuals (pp. 96–101), and also to an article in the *Gardeners' Chronicle* a year and a half ago, for the details of a very interesting contribution to science, irrespective of theory.

In domestication, this intercrossing may be prevented; and in this prevention lies the art of producing varieties. But "the art itself is Nature," since the whole art consists in allowing the most universal of all natural tendencies in organic things (inheritance) to operate uncontrolled by other and obviously incidental tendencies. No new power, no artificial force, is brought into play either by separating the stock of a desirable variety so as to prevent mixture, or by selecting for breeders those individuals which most largely partake of the pecularities for which the breed is valued.[1]

We see everywhere around us the remarkable results which Nature may be said to have brought about under artificial selection and separation. Could she accomplish similar results when left to herself? Variations might begin, we know they do begin, in a wild state. But would any of them be preserved and carried to an equal degree of deviation? Is there anything in Nature which in the long-run may answer to

[1] The rules and processes of breeders of animals, and their results, are so familiar that they need not be particularized. Less is popularly known about the production of vegetable races. We refer our readers back to this Journal, vol. xxvii., pp. 440–442 (May, 1859), for an abstract of the papers of M. Vilmorin upon this subject.

artificial selection? Mr. Darwin thinks that there is; and *Natural Selection* is the key-note of his discourse.

As a preliminary, he has a short chapter to show that there is variation in Nature, and therefore something for natural selection to act upon. He readily shows that such mere variations as may be directly referred to physical conditions (like the depauperation of plants in a sterile soil, or their dwarfing as they approach an Alpine summit, the thicker fur of an animal from far northward, etc.), and also those individual differences which we everywhere recognize but do not pretend to account for, are not separable by any assignable line from more strongly-marked varieties; likewise that there is no clear demarkation between the latter and sub-species, or varieties of the higest grade (distinguished from species not by any known inconstancy, but by the supposed lower importance of their characteristics); nor between these and recognized species. "These differences blend into each other in an insensible series, and the series impresses the mind with an idea of an actual passage."

This gradation from species downward is well made out. To carry it one step farther upward, our author presents in a strong light the differences which prevail among naturalists as to what forms should be admitted to the rank of species. Some genera (and these in some countries) give rise to far more discrepancy than others; and it is concluded that the large or dominant genera are usually the most variable. In a flora so small as the British, 182 plants, generally reckoned as varieties, have been ranked by some botanists as species. Selecting the British genera which include the most polymorphous forms, it appears that Babington's Flora gives them 251 species, Bentham's only 112, a difference of 139 doubtful forms. These are nearly the extreme views, but they are the views of two most capable and most experienced judges, in respect to one of the best-known floras of the world. The fact is suggestive, that the best-known countries furnish the greatest number of such doubtful cases. Illustrations of this kind may be multiplied to a great extent. They make it plain that, whether species in Nature are aboriginal and definite or not, our practical conclusions about them, as embodied in systematic works, are not *facts* but *judgments*, and largely fallible judgments.

How much of the actual coincidence of authorities is owing to imperfect or restricted observation, and to one naturalist's adopting the conclusions of another without independent observation, this is not the place to consider. It is our impression that species of animals are more definitely marked than those of plants; this may arise from our somewhat extended acquaintance with the latter, and our ignorance of the former. But we are constrained by our experience to admit the strong likelihood, in botany, that varieties on the one hand, and what are called closely-related species on the other, do not differ except in degree. Whenever this wider difference separating the latter can be spanned by intermediate forms, as it sometimes is, no botanist long resists the inevitable conclusion. Whenever, therefore, this wider difference can be shown to be compatible with community of origin, and explained through natural selection or in any other way, we are ready to adopt the *probable* conclusion; and we see beforehand how strikingly the actual geographical association of related species favors the broader view. Whether we should continue to regard the forms in question as distinct species, depends upon what meaning we shall finally attach to that term; and that depends upon how far the doctrine of derivation can be carried back and how well it can be supported.

In applying his principle of natural selection to the work in hand, Mr. Darwin assumes, as we have seen: 1. Some variability of animals and plants in nature; 2. The absence of any definite distinction between slight variations, and varieties of the highest grade; 3. The fact that naturalists do not practically agree, and do not increasingly tend to agree, as to what forms are species and what are strong varieties, thus rendering it probable that there may be no essential and original difference, or no possibility of ascertaining it, at least in many cases; also, 4. That the most flourishing and dominant species of the larger genera on an average vary most (a proposition which can be substantiated only by extensive comparisons, the details of which are not given); and, 5. That in large genera the species are apt to be closely but unequally allied together, forming little clusters round certain species—just such clusters as would be formed if we suppose their members once to have been satellites or varieties of a central or parent species, but to have attained at length a wider divergence and a specific character. The fact of such association is undeniable; and the use which Mr. Darwin makes of it seems fair and natural.

The gist of Mr. Darwin's work is to show that such varieties are gradually diverged into species and genera through *natural selection;* that natural selection is the inevitable result of the *struggle for existence* which all living things are engaged in; and that this struggle is an unavoidable consequence of several natural causes, but mainly of the high rate at which all organic beings tend to increase.

Curiously enough, Mr. Darwin's theory is grounded upon the doctrine of Malthus and the doctrine of Hobbes. The elder DeCandolle had conceived the idea of the struggle for existence, and, in a passage which would have delighted the cynical philosopher of Malmesbury, had declared that all Nature is at war, one organism with another or with external Nature; and Lyell and Herbert had made considerable use of it. But Hobbes in his theory of society, and Darwin in his theory of natural history, alone have built their systems upon it. However moralists and political economists may regard these doctrines in their original application to human society and the relation of population to subsistence, their thorough applicability to the great society of the organic world in general is now undeniable. And to Mr. Darwin belongs the credit of making this extended application, and of working out the immensely diversified results with rare sagacity and untiring patience. He has brought to view *real causes* which have been largely operative in the establishment of the actual association and geographical distribution of plants and animals. In this he must be allowed to have made a very important contribution to an interesting department of science, even if his theory fails in the endeavor to explain the origin or diversity of species.

"Nothing is easier," says our author, "than to admit in words the truth of the universal struggle for life, or more difficult—at least I have found it so—than constantly to bear this conclusion in mind. Yet, unless it be thoroughly ingrained in the mind, I am convinced that the whole economy of Nature, with every fact on distribution, rarity, abundance, extinction, and variation, will be dimly seen or quite misunderstood. We behold the face of Nature bright with gladness, we often see superabundance of food; we do not see, or we forget, that the birds which are idly singing round us mostly live on insects or seeds, and are thus constantly destroying life; or we forget how largely these songsters, or their eggs, or their nestlings, are destroyed by birds and beasts of prey; we do not always bear in mind that, though food may be now superabundant, it is not so at all seasons of each recurring year."—(p. 62.)

"There is no exception to the rule that every organic being naturally increases at so high a rate that, if not destroyed, the earth would soon be covered by the progeny of a single pair. Even slow-breeding man has doubled in twenty-five years, and at this rate, in a few thousand years, there would literally not be standing-room for his progeny. Linnæus has calculated that if an annual plant produced only two seeds—and there is no plant so unproductive as this—and their seedlings next year produced two, and so on, then in twenty years there would be a million plants. The elephant is reckoned to be the slowest breeder of all known animals, and I have taken some pains to estimate its probable minimum rate of natural increase; it will be under the mark to assume that it breeds when thirty years old, and goes on breeding till ninety years old, bringing forth three pairs of young in this interval; if this be so, at the end of the fifth century there would be alive fifteen million elephants, descended from the first pair.

"But we have better evidence on this subject than mere theoretical calculations, namely, the numerous recorded cases of the astonishingly rapid increase of various animals in a state of nature, when circumstances have been favorable to them during two or three following seasons. Still more striking is the evidence from our domestic animals of many kinds which have run wild in several parts of the world; if the statements of the rate of increase of slow-breeding cattle and horses in South America, and latterly in Australia, had not been well authenticated, they would have been quite incredible. So it is with plants: cases could be given of introduced plants which have become common throughout whole islands in a period of less than ten years. Several of the plants now most numerous over the wide plains of La Plata, clothing square leagues of surface almost to the exclusion of all other plants, have been introduced from Europe; and there are plants which now range in India, as I hear from Dr. Falconer, from Cape Comorin to the Himalaya, which have been imported from America since its discovery. In such cases, and endless instances could be given, no one supposes that the fertility of these animals or plants has been suddenly and temporarily increased in any sensible degree. The obvious explanation is, that the conditions of life have been very favorable, and that there has consequently been less destruction of the old and young, and that nearly all the young have been enabled to breed. In such cases the geometrical ratio of increase, the result of which never fails to be surprising, simply explains the extraordinarily rapid increase and wide diffusion of naturalized productions in their new homes."—(pp. 64, 65.)

"All plants and animals are tending to increase at a geometrical ratio; all would most rapidly stock any station in which they could anyhow exist; the increase must be checked by destruction at some period of life."—(p. 65.)

The difference between the most and the least prolific species is of no account:

"The condor lays a couple of eggs, and the ostrich a score; and yet in the same country the condor may be the more numerous of the two. The Fulmar petrel lays but one egg, yet it is believed to be the most numerous bird in the world."—(p. 68.)

"The amount of food gives the extreme limit to which each

species can increase; but very frequently it is not the obtaining of food, but the serving as prey to other animals, which determines the average numbers of species."—(p. 68.)

"Climate plays an important part in determining the average numbers of a species, and periodical seasons of extreme cold or drought I believe to be the most effective of all checks. I estimated that the winter of 1854–'55 destroyed four-fifths of the birds in my own grounds; and this is a tremendous destruction, when we remember that ten per cent. is an extraordinarily severe mortality from epidemics with man. The action of climate seems at first sight to be quite independent of the struggle for existence; but, in so far as climate chiefly acts in reducing food, it brings on the most severe struggle between the individuals, whether of the same or of distinct species, which subsist on the same kind of food. Even when climate, for instance extreme cold, acts directly, it will be the least vigorous, or those which have got least food through the advancing winter, which will suffer most. When we travel from south to north, or from a damp region to a dry, we invariably see some species gradually getting rarer and rarer, and finally disappearing; and, the change of climate being conspicuous, we are tempted to attribute the whole effect to its direct action. But this is a very false view; we forget that each species, even where it most abounds, is constantly suffering enormous destruction at some period of its life, from enemies or from competitors for the same place and food; and if these enemies or competitors be in the least degree favored by any slight change of climate, they will increase in numbers, and, as each area is already stocked with inhabitants, the other species will decrease. When we travel southward and see a species decreasing in numbers, we may feel sure that the cause lies quite as much in other species being favored as in this one being hurt. So it is when we travel northward, but in a somewhat lesser degree, for the number or species of all kinds, and therefore of competitors, decreases northward; hence, in going northward, or in ascending a mountain, we far oftener meet with stunted forms, due to the *directly* injurious action of climate, than we do in proceeding southward or in descending a mountain. When we reach the arctic regions, or snow-capped summits, or absolute deserts, the struggle for life is almost exclusively with the elements.

"That climate acts in main part indirectly by favoring other species, we may clearly see in the prodigious number of plants in our gardens which can perfectly well endure our climate, but which never become naturalized, for they cannot compete with our native plants, nor resist destruction by our native animals." —(pp. 68, 69.)

After an instructive instance in which "cattle absolutely determine the existence of the Scotch fir," we are referred to cases in which insects determine the existence of cattle:

"Perhaps Paraguay offers the most curious instance of this; for here neither cattle, nor horses, nor dogs, have ever run wild, though they swarm southward and northward in a feral state; and Azara and Rengger have shown that this is caused by the greater number in Paraguay of a certain fly, which lays its eggs in the navels of these animals when first born. The increase of these flies, numerous as they are, must be habitually checked by some means, probably by birds. Hence, if certain insectivorous birds (whose numbers are probably regulated by hawks or beasts of prey) were to increase in Paraguay, the flies would decrease—then cattle and horses would become feral, and this would certainly greatly alter (as indeed I have observed in parts of South America) the vegetation; this, again, would, largely affect the insects; and this, as we have just seen in Staffordshire, the insectivorous birds, and so onward in ever-increasing circles of complexity. We began this series by insectivorous birds, and we had ended with them. Not that in Nature the relations can ever be as simple as this. Battle within battle must ever be recurring with varying success; and yet in the long-run the forces are so nicely balanced that the face of Nature remains uniform for long periods of time, though assuredly the merest trifle would often give the victory to one organic being over another. Nevertheless, so profound is our ignorance, and so high our presumption, that we marvel when we hear of the extinction of an organic being; and as we do not see the cause, we invoke cataclysms to desolate the world, or invent laws on the duration of the forms of life!"—(pp. 72, 73.)

"When we look at the plants and bushes clothing an entangled bank, we are tempted to attribute their proportional numbers and kinds to what we call chance. But how false a view is this! Every one has heard that when an American forest is cut down, a very different vegetation springs up; but it has been observed that the trees now growing on the ancient Indian mounds, in the Southern United States, display the same beautiful diversity and proportion of kinds as in the surrounding virgin forests. What a struggle between the several kinds of trees must here have gone on during long centuries, each annually scattering its seeds by the thousand; what war between insect and insect—between insects, snails, and other animals, with birds and beasts of prey—all striving to increase, and all feeding on each other or on the trees, or their seeds and seedlings, or on the other plants which first clothed the ground and thus checked the growth of the trees! Throw up a handful of feathers, and all must fall to the ground according to definite laws; but how simple is this problem compared to the action and reaction of the innumerable plants and animals which have determined, in the course of centuries, the proportional numbers and kinds of trees now growing on the old Indian ruins!"—(pp. 74, 75.)

For reasons obvious upon reflection, the competition is often, if not generally, most severe betwen nearly related species when they are in contact, so that one drives the other before it, as the Hanoverian the old English rat, the small Asiatic cockroach in Russia, its greater congener, etc. And this, when duly considered, explains many curious results; such, for instance, as the considerable number of different genera of plants and animals which are generally found to inhabit any limited area.

"The truth of the principle that the greatest amount of life can be supported by great diversification of structure is seen under many natural circumstances. In an extremely small area, especially if freely open to immigration, and where the contest between individual and individual must be severe, we always find great diversity in its inhabitants. For instance, I found that a piece of turf, three feet by four in size, which had been exposed for many years to exactly the same conditions, supported twenty species of plants, and these belonged to eighteen genera, and to eight orders, which showed how much these plants differed from each other. So it is with the plants and insects on small and uniform islets; and so in small ponds of fresh water. Farmers find that they can raise most food by a rotation of plants belonging to the most different orders; Nature follows what may be called a simultaneous rotation. Most of the animals and plants which live close round any small piece of ground could live on it (supposing it not to be in any way peculiar in its nature), and may be said to be striving to the utmost to live there; but it is seen that, where they come into the closest competition with each other, the advantages of diversification of structure, with the accompanying differences of habit and constitution, determine that the inhabitants, which thus jostle each other most closely, shall, as a general rule, belong to what we call different genera and orders."—(p. 114.)

The abundance of some forms, the rarity and final extinction of many others, and the consequent divergence of character or increase of difference among the surviving representatives, are other consequences. As favored forms increase, the less favored must diminish in number, for there is not room for all; and the slightest advantage, at first probably inappreciable to human observation, must decide which shall prevail and which must perish, or be driven to another and for it more favorable locality.

We cannot do justice to the interesting chapter

upon natural selection by separated extracts. The following must serve to show how the principle is supposed to work:

"If during the long course of ages, and under varying conditions of life, organic beings vary at all in the several parts of their organization, and I think this cannot be disputed; if there be, owing to the high geometrical powers of increase of each species, at some age, season, or year, a severe struggle for life, and this certainly cannot be disputed: then, considering the infinite complexity of the relations of all organic beings to each other and to their conditions of existence, causing an infinite diversity in structure, constitution, and habits, to be advantageous to them, I think it would be a most extraordinary fact if no variation ever had occurred useful to each being's own welfare, in the same way as so many variations have occurred useful to man. But if variations useful to any organic being do occur, assuredly individuals thus characterized will have the best chance of being preserved in the struggle for life; and from the strong principle of inheritance they will tend to produce offspring similarly characterized. This principle of preservation I have called, for the sake of brevity, Natural Selection."—(pp. 126, 127.)

"In order to make it clear how, as I believe, natural selection acts, I must beg permission to give one or two imaginary illustrations. Let us take the case of a wolf, which preys on various animals, securing some by craft, some by strength, and some by fleetness; and let us suppose that the fleetest prey, a deer for instance, had from any change in the country increased in numbers, or that other prey had decreased in numbers, during that season of the year when the wolf is hardest pressed for food. I can under such circumstances see no reason to doubt that the swiftest and slimmest wolves would have the best chance of surviving, and so be preserved or selected—provided always that they retained strength to master their prey at this or at some other period of the year, when they might be compelled to prey on other animals. I can see no more reason to doubt this than that man can improve the fleetness of his greyhounds by careful and methodical selection, or by that unconscious selection which results from each man trying to keep the best dogs without any thought of modifying the breed.

"Even without any change in the proportional numbers of the animals on which our wolf preyed, a cub might be born with an innate tendency to pursue certain kinds of prey. Nor can this be thought very improbable; for we often observe great differences in the natural tendencies of our domestic animals: one cat, for instance, taking to catching rats, another mice; one cat, according to Mr. St. John, bringing home winged game, another hares or rabbits, and another hunting on marshy ground, and almost nightly catching woodcocks or snipes. The tendency to catch rats rather than mice is known to be inherited. Now, if any slight innate change of habit or of structure benefited an individual wolf, it would have the best chance of surviving and of leaving offspring. Some of its young would probably inherit the same habits or structure, and by the repetition of this process a new variety might be formed which would either supplant or coexist with the parent-form of wolf. Or, again, the wolves inhabiting a mountainous district, and those frequenting the lowlands, would naturally be forced to hunt different prey; and from a continued preservation of the individuals best fitted for the two sites, two varieties might slowly be formed. These varieties would cross and blend where they met; but to this subject of intercrossing we shall soon have to return. I may add that, according to Mr. Pierce, there are two varieties of the wolf inhabiting the Catskill Mountains in the United States, one with a light greyhound-like form, which pursues deer, and the other more bulky, with shorter legs, which more frequently attacks the shepherd's flock."—(pp. 90, 91.)

We eke out the illustration here with a counterpart instance, viz., the remark of Dr. Bachman that "the deer that reside permanently in the swamps of Carolina are taller and longer-legged than those in the higher grounds."[1]

[1] "Quadrupeds of America," vol. ii., p. 239.

The limits allotted to this article are nearly reached, yet only four of the fourteen chapters of the volume have been touched. These, however, contain the fundamental principles of the theory, and most of those applications of it which are capable of something like verification, relating as they do to the phenomena now occurring. Some of our extracts also show how these principles are thought to have operated through the long lapse of the ages. The chapters from the sixth to the ninth inclusive are designed to obviate difficulties and objections, "some of them so grave that to this day," the author frankly says, he "can never reflect on them without being staggered." We do not wonder at it. After drawing what comfort he can from "the imperfection of the geological record" (Chapter IX.), which we suspect is scarcely exaggerated, the author considers the geological succession of organic beings (Chapter X.), to see whether they better accord with the common view of the immutability of species, or with that of their slow and gradual modification. Geologists must settle that question. Then follow two most interesting and able chapters on the geographical distribution of plants and animals, the summary of which we should be glad to cite; then a fitting chapter upon classification, morphology, embryology, etc., as viewed in the light of this theory, closes the argument; the fourteenth chapter being a recapitulation.

The interest for the general reader heightens as the author advances on his perilous way and grapples manfully with the most formidable difficulties.

To account, upon these principles, for the gradual

elimination and segregation of nearly allied forms—such as varieties, sub-species, and closely-related or representative species—also in a general way for their geographical association and present range, is comparatively easy, is apparently within the bounds of possibility. Could we stop here we should be fairly contented. But, to complete the system, to carry out the principles to their ultimate conclusion, and to explain by them many facts in geographical distribution which would still remain anomalous, Mr. Darwin is equally bound to account for the formation of genera, families, orders, and even classes, by natural selection. He does "not doubt that the theory of descent with modification embraces all the members of the same class," and he concedes that analogy would press the conclusion still further; while he admits that "the more distinct the forms are, the more the arguments fall away in force." To command assent we naturally require decreasing probability to be overbalanced by an increased weight of evidence. An opponent might plausibly, and perhaps quite fairly, urge that the links in the chain of argument are weakest just where the greatest stress falls upon them.

To which Mr. Darwin's answer is, that the best parts of the testimony have been lost. He is confident that intermediate forms must have existed; that in the olden times when the genera, the families, and the orders, diverged from their parent stocks, gradations existed as fine as those which now connect closely related species with varieties. But they have passed and left no sign. The geological record, even if all displayed to view, is a book from which not only many pages, but even whole alternate chapters, have been lost out, or rather which were never printed from the autographs of Nature. The record was actually made in fossil lithography only at certain times and under certain conditions (i. e., at periods of slow subsidence and places of abundant sediment); and of these records all but the last volume is out of print; and of its pages only local glimpses have been obtained. Geologists, except Lyell, will object to this—some of them moderately, others with vehemence. Mr. Darwin himself admits, with a candor rarely displayed on such occasions, that he should have expected more geological evidence of transition than he finds, and that all the most eminent paleontologists maintain the immutability of species.

The general fact, however, that the fossil fauna of each period as a whole is nearly intermediate in character between the preceding and the succeeding faunas, is much relied on. We are brought one step nearer to the desired inference by the similar "fact, insisted on by all paleontologists, that fossils from two consecutive formations are far more closely related to each other than are the fossils of two remote formations. Pictet gives a well-known instance—the general resemblance of the organic remains from the several stages of the chalk formation, though the species are distinct at each stage. This fact alone, from its generality, seems to have shaken Prof. Pictet in his firm belief in the immutability of species" (p. 335). What Mr. Darwin now particularly wants to complete his inferential evidence is a proof that the same gradation may be traced in later periods, say in the Tertiary, and between that period and the present; also that the later gradations are finer, so as to leave it doubtful whether the succession is one of species—believed on the one theory to be independent, on the other, derivative—or of varieties, which are confessedly derivative. The proof of the finer gradation appears to be forthcoming. Des Hayes and Lyell have concluded that many of the middle Tertiary and a large proportion of the later Tertiary mollusca are specifically identical with living species; and this is still the almost universally prevalent view. But Mr. Agassiz states that, "in every instance where he had sufficient materials, he had found that the species of the two epochs supposed to be identical by Des Hayes and Lyell were in reality distinct, although closely allied species." [1] Moreover, he is now satisfied, as we understand, that the same gradation is traceable not merely in each great division of the Tertiary, but in particular deposits or successive beds, each answering to a great number of years; where what have passed unquestioned as members of one species, upon closer examination of numerous specimens exhibit differences which in his opinion entitle them to be distinguished into two, three, or more species. It is plain, therefore, that whatever conclusions can be fairly drawn from the present animal and vegetable kingdoms in favor of a gradation of varieties into species, or into what may be regarded as such, the same may be extended to the Tertiary period. In both cases, what some call species others call varieties; and in the later Tertiary shells

[1] "Proceedings of the American Academy of Arts and Sciences," vol. iv., p. 178.

this difference in judgment affects almost half of the species!

We pass to a second difficulty in the way of Mr. Darwin's theory; to a case where we are perhaps entitled to demand of him evidence of gradation like that which connects the present with the Tertiary mollusca. Wide, very wide is the gap, anatomically and physiologically (we do not speak of the intellectual) between the highest quadrumana and man; and comparatively recent, if ever, must the line have bifurcated. But where is there the slightest evidence of a common progenitor? Perhaps Mr. Darwin would reply by another question: where are the fossil remains of the men who made the flint knives and arrowheads of the Somme Valley?

We have a third objection, one, fortunately, which has nothing to do with geology. We can only state it here in brief terms. The chapter on hybridism is most ingenious, able, and instructive. If sterility of crosses is a special original arrangement to prevent the confusion of species by mingling, as is generally assumed, then, since varieties cross readily and their offspring is fertile *inter se*, there is a fundamental distinction between varieties and species. Mr. Darwin therefore labors to show that it is not a special endowment, but an incidental acquirement. He does show that the sterility of crosses is of all degrees; upon which we have only to say, *Natura non facit saltum*, here any more than elsewhere. But, upon his theory he is bound to show how sterility might be acquired, through natural selection or through something else. And the difficulty is, that, whereas individuals of the very same blood tend to be sterile, and somewhat remoter unions diminish this tendency, and when they have diverged into two varieties the cross-breeds between the two are more fertile than either pure stock —yet when they have diverged only one degree more the whole tendency is reversed, and the mongrel is sterile, either absolutely or relatively. He who explains the genesis of species through purely natural agencies should assign a natural cause for this remarkable resült; and this Mr. Darwin has not done. Whether original or derived, however, this arrangement to keep apart those forms which have, or have acquired (as the case may be), a certain moderate amount of difference, looks to us as much designed for the purpose, as does a ratchet to prevent reverse motion in a wheel. If species have originated by divergence, this keeps them apart.

Here let us suggest a possibly attainable test of the theory of derivation, a kind of instance which Mr. Darwin may be fairly asked to produce—viz., an instance of two varieties, or what may be assumed as such, which have diverged enough to reverse the movement, to bring out some sterility in the crosses. The best marked human races might offer the most likely case. If mülattoes are sterile or tend to sterility, as some naturalists confidently assert, they afford Mr. Darwin a case in point. If, as others think, no such tendency is made out, the required evidence is wanting.

A fourth and the most formidable difficulty is that of the production and specialization of organs.

It is well said that all organic beings have been formed on two great laws: unity of type, and adaptation to the conditions of existence.[1] The special teleologists, such as Paley, occupy themselves with the latter only; they refer particular facts to special design, but leave an overwhelming array of the widest facts inexplicable. The morphologists build on unity of type, or that fundamental agreement in the structure of each great class of beings which is quite independent of their habits or conditions of life; which requires each individual "to go through a certain formality," and to accept, at least for a time, certain organs, whether they are of any use to him or not. Philosophical minds form various conceptions for harmonizing the two views theoretically. Mr. Darwin harmonizes and explains them naturally. Adaptation to the conditions of existence is the result of natural selection; unity of type, of unity of descent. Accordingly, as he puts his theory, he is bound to account for the origination of new organs, and for their diversity in each great type, for their specialization, and every adaptation of organ to function and of structure to condition, through natural agencies. Whenever he attempts this he reminds us of Lamarck, and shows us how little light the science of a century devoted to structural investigation has thrown upon the mystery of organization. Here purely natural explanations fail. The organs being given, natural selection may account for some improvement; if given of a variety of sorts or grades, natural selection might determine which should survive and where it should prevail.

On all this ground the only line for the theory to

[1] Owen adds a third, viz., vegetative repetition; but this, in the vegetable kingdom, is simply unity of type.

take is to make the most of gradation and adherence to type as suggestive of derivation, and unaccountable upon any other scientific view—deferring all attempts to explain *how* such a metamorphosis was effected, until naturalists have explained *how* the tadpole is metamorphosed into a frog, or one sort of polyp into another. As to *why* it is so, the philosophy of efficient cause, and even the whole argument from design, would stand, upon the admission of such a theory of derivation, precisely where they stand without it. At least there is, or need be, no ground of difference here between Darwin and Agassiz. The latter will admit, with Owen and every morphologist, that hopeless is the attempt to explain the similarity of pattern in members of the same class by utility or the doctrine of final causes. "On the ordinary view of the independent creation of each being, we can only say that so it is, that it has so pleased the Creator to construct each animal and plant." Mr. Darwin, in proposing a theory which suggests a *how* that harmonizes these facts into a system, we trust implies that all was done wisely, in the largest sense designedly, and by an intelligent first cause. The contemplation of the subject on the intellectual side, the amplest exposition of the unity of plan in creation, considered irrespective of natural agencies, leads to no other conclusion.

We are thus, at last, brought to the question, What would happen if the derivation of species were to be substantiated, either as a true physical theory, or as a sufficient hypothesis? What would come of it? The inquiry is a pertinent one, just now. For, of those who agree with us in thinking that Darwin has not established his theory of derivation many will admit with us that he has rendered a theory of derivation much less improbable than before; that such a theory chimes in with the established doctrines of physical science, and is not unlikely to be largely accepted long before it can be proved. Moreover, the various notions that prevail—equally among the most and the least religious—as to the relations between natural agencies or phenomena and efficient cause, are seemingly more crude, obscure, and discordant, than they need be.

It is not surprising that the doctrine of the book should be denounced as atheistical. What does surprise and concern us is, that it should be so denounced by a scientific man, on the broad assumption that a material connection between the members of a series of organized beings is inconsistent with the idea of their being intellectually connected with one another through the Deity, i. e., as products of one mind, as indicating and realizing a preconceived plan. An assumption the rebound of which is somewhat fearful to contemplate, but fortunately one which every natural birth protests against.

It would be more correct to say that the theory in itself is perfectly compatible with an atheistic view of the universe. That is true; but it is equally true of physical theories generally. Indeed, it is more true of the theory of gravitation, and of the nebular hypothesis, than of the hypothesis in question. The latter merely takes up *a particular, proximate cause*, or set of such causes, from which, it is argued, the present diversity of species has or may have *contingently* resulted. The author does not say *necessarily* resulted;

that the actual results in mode and measure, and none other, must have taken place. On the other hand, the theory of gravitation and its extension in the nebular hypothesis assume a *universal and ultimate* physical cause, from which the effects in Nature must *necessarily* have resulted. Now, it is not thought, at least at the present day, that the establishment of the Newtonian theory was a step toward atheism or pantheism. Yet the great achievement of Newton consisted in proving that certain forces (blind forces, so far as the theory is concerned), acting upon matter in certain directions, must *necessarily* produce planetary orbits of the exact measure and form in which observation shows them to exist—a view which is just as consistent with eternal necessity, either in the atheistic or the pantheistic form, as it is with theism.

Nor is the theory of derivation particularly exposed to the charge of the atheism of fortuity; since it undertakes to assign real causes for harmonious and systematic results. But, of this, a word at the close.

The value of such objections to the theory of derivation may be tested by one or two analogous cases. The common scientific as well as popular belief is that of the original, independent creation of oxygen and hydrogen, iron, gold, and the like. Is the speculative opinion now increasingly held, that some or all of the supposed elementary bodies are derivative or compound, developed from some preceding forms of matter, irreligious? Were the old alchemists atheists as well as dreamers in their attempts to transmute earth into gold? Or, to take an instance from force (power) —which stands one step nearer to efficient cause than form—was the attempt to prove that heat, light, electricity, magnetism, and even mechanical power, are variations or transmutations of one force, atheistical in its tendency? The supposed establishment of this view is reckoned as one of the greatest scientific triumphs of this century.

Perhaps, however, the objection is brought, not so much against the speculation itself, as against the attempt to show how derivation might have been brought about. Then the same objection applies to a recent ingenious hypothesis made to account for the genesis of the chemical elements out of the ethereal medium, and to explain their several atomic weights and some other characteristics by their successive complexity—hydrogen consisting of so many atoms of ethereal substance united in a particular order, and so on. The speculation interested the philosophers of the British Association, and was thought innocent, but unsupported by facts. Surely Mr. Darwin's theory is none the worse, morally, for having some foundation in fact.

In our opinion, then, it is far easier to vindicate a theistic character for the derivative theory, than to establish the theory itself upon adequate scientific evidence. Perhaps scarcely any philosophical objection can be urged against the former to which the nebular hypothesis is not equally exposed. Yet the nebular hypothesis finds general scientific acceptance, and is adopted as the basis of an extended and recondite illustration in Mr. Agassiz's great work.[1]

How the author of this book harmonizes his scientific theory with his philosophy and theology, he has

[1] "Contributions to Natural History of America," vol. i., pp. 127-131.

not informed us. Paley in his celebrated analogy with the watch, insists that if the timepiece were so constructed as to produce other similar watches, after a manner of generation in animals, the argument from design would be all the stronger. What is to hinder Mr. Darwin from giving Paley's argument a further *a-fortiori* extension to the supposed case of a watch which sometimes produces better watches, and contrivances adapted to successive conditions, and so at length turns out a chronometer, a town clock, or a series of organisms of the same type? From certain incidental expressions at the close of the volume, taken in connection with the motto adopted from Whewell, we judge it probable that our author regards the whole system of Nature as one which had received at its first formation the impress of the will of its Author, foreseeing the varied yet necessary laws of its action throughout the whole of its existence, ordaining when and how each particular of the stupendous plan should be realized in effect, and—with Him to whom to will is to do—in ordaining doing it. Whether profoundly philosophical or not, a view maintained by eminent philosophical physicists and theologians, such as Babbage on the one hand and Jowett on the other, will hardly be denounced as atheism. Perhaps Mr. Darwin would prefer to express his idea in a more general way, by adopting the thoughtful words of one of the most eminent naturalists of this or any age, substituting the word *action* for "thought," since it is the former (from which alone the latter can be inferred) that he has been considering. "Taking Nature as exhibiting thought for my guide, it appears to me that while human thought is consecutive, Divine thought is simultaneous, embracing at the same time and forever, in the past, the present and the future, the most diversified relations among hundreds of thousands of organized beings, each of which may present complications again, which to study and understand even imperfectly—as for instance man himself—mankind has already spent thousands of years."[1] In thus conceiving of the Divine Power in act as coetaneous with Divine Thought, and of both as far as may be apart from the human element of time, our author may regard the intervention of the Creator either as, humanly speaking, *done from all time*, or else as *doing through all time*. In the ultimate analysis we suppose that every philosophical theist must adopt one or the other conception.

A perversion of the first view leads toward atheism, the notion of an eternal sequence of cause and effect, for which there is no first cause—a view which few sane persons can long rest in. The danger which may threaten the second view is pantheism. We feel safe from either error, in our profound conviction that there is order in the universe; that order presupposes mind; design, will; and mind or will, personality. Thus guarded, we much prefer the second of the two conceptions of causation, as the more philosophical as well as Christian view—a view which leaves us with the same difficulties and the same mysteries in Nature as in Providence, and no other. Natural law, upon this view, is the human conception of continued and orderly Divine action.

[1] *Op. cit.*, p. 130.

We do not suppose that less power, or other power, is required to sustain the universe and carry on its operations, than to bring it into being. So, while conceiving no improbability of "interventions of Creative mind in Nature," if by such is meant the bringing to pass of new and fitting events at fitting times, we leave it for profounder minds to establish, if they can, a rational distinction in kind between his working in Nature carrying on operations, and in initiating those operations.

We wished, under the light of such views, to examine more critically the doctrine of this book, especially of some questionable parts; for instance, its explanation of the natural development of organs, and its implication of a "necessary acquirement of mental power" in the ascending scale of gradation. But there is room only for the general declaration that we cannot think the Cosmos a series which began with chaos and ends with mind, or of which mind is a result: that, if, by the successive origination of species and organs through natural agencies, the author means a series of events which succeed each other irrespective of a continued directing intelligence—events which mind does not order and shape to destined ends—then he has not established that doctrine, nor advanced toward its establishment, but has accumulated improbabilities beyond all belief. Take the formation and the origination of the successive degrees of complexity of eyes as a specimen. The treatment of this subject (pp. 188, 189), upon one interpretation, is open to all the objections referred to; but, if, on the other hand, we may rightly compare the eye "to a telescope, perfected by the long-continued efforts of the highest human intellects," we could carry out the analogy, and draw satisfactory illustrations and inferences from it. The essential, the directly intellectual thing is the making of the improvements in the telescope or the steam-engine. Whether the successive improvements, being small at each step, and consistent with the general type of the instrument, are applied to some of the individual machines, or entire new machines are constructed for each, is a minor matter. Though, if machines could engender, the adaptive method would be most economical; and economy is said to be a paramount law in Nature. The origination of the improvements, and the successive adaptations to meet new conditions or subserve other ends, are what answer to the supernatural, and therefore remain inexplicable. As to bringing them into use, though wisdom foresees the result, the circumstances and the natural competition will take care of that, in the long-run. The old ones will go out of use fast enough, except where an old and simple machine remains still best adapted to a particular purpose or condition—as, for instance, the old Newcomen engine for pumping out coal-pits. If there's a Divinity that shapes these ends, the whole is intelligible and reasonable; otherwise, not.

We regret that the necessity of discussing philosophical questions has prevented a fuller examination of the theory itself, and of the interesting scientific points which are brought to bear in its favor. One of its neatest points, certainly a very strong one for the local origination of species, and their gradual diffusion under natural agencies, we must reserve for some other convenient opportunity.

The work is a scientific one, rigidly restricted to its direct object; and by its science it must stand or fall. Its aim is, probably, not to deny creative intervention in Nature—for the admission of the independent origination of certain types does away with all antecedent improbability of as much intervention as may be required—but to maintain that Natural Selection, in explaining the facts, explains also many classes of facts which thousand-fold repeated independent acts of creation do not explain, but leave more mysterious than ever. How far the author has succeeded, the scientific world will in due time be able to pronounce.

As these sheets are passing through the press, a copy of the second edition has reached us. We notice with pleasure the insertion of an additional motto on the reverse of the title-page, directly claiming the theistic view which we have vindicated for the doctrine. Indeed, these pertinent words of the eminently wise Bishop Butler comprise, in their simplest expression, the whole substance of our later pages:

"The only distinct meaning of the word 'natural' is *stated*, *fixed*, or *settled*; since what is natural as much requires and presupposes an intelligent mind to render it so, i. e., to effect it continually or at stated times, as what is supernatural or miraculous does to effect it for once."

II.

DESIGN VERSUS NECESSITY.—DISCUSSION BETWEEN TWO READERS OF DARWIN'S TREATISE ON THE ORIGIN OF SPECIES, UPON ITS NATURAL THEOLOGY.

(American Journal of Science and Arts, *September*, 1860.)

D. T.—Is Darwin's theory atheistic or pantheistic? or, does it tend to atheism or pantheism? Before attempting any solution of this question, permit me to say a few words tending to obtain a definite conception of *necessity* and *design*, as the sources from which events may originate, each independent of the other; and we shall, perhaps, best attain a clear understanding of each, by the illustration of an example in which simple human designers act upon the physical powers of common matter.

Suppose, then, a square billiard-table to be placed with its corners directed to the four cardinal points. Suppose a player, standing at the north corner, to strike a red ball directly to the south, his design being to lodge the ball in the south pocket; which design, if not interfered with, must, of course be accomplished. Then suppose another player, standing at the east corner, to direct a white ball to the west corner. This design also, if not interfered with, must be accomplished. Next suppose both players to strike their balls at the same instant, with like forces, in the directions before given. In this case the balls would not pass as before, namely, the red ball to the south, and the white ball to the west, but they must both meet and strike each other in the centre of the table, and, being perfectly elastic, the red ball must pass to the west pocket, and the white ball to the south pocket. We may suppose that the players acted wholly without concert with each other, indeed, they may be ignorant of each other's design, or even of each other's existence; still we know that the events must happen as herein described. Now, the first half of the course of these two balls is from an impulse, or proceeds from a power, acting from design. Each player has the design of driving his ball across the table in a diagonal line to accomplish its lodgment at the opposite corner of the table. Neither designed that his ball should be deflected from that course and pass to another corner of the table. The *direction* of this second part of the motion must be referred entirely to *necessity*, which directly interferes with the purpose of him who designed the rectilinear direction. We are not, in this case, to go back to find design in the creation of the powers or laws of inertia and elasticity, after the order of which the deflection, at the instant of collision, necessarily takes place. We know that these powers were inherent in the balls, and were not created to answer this special deflection. We are required, by the hypothesis, to confine attention in point of time, from the instant preceding the impact of the balls, to the time of their arrival at the opposite corners of the table. The cues are moved by design. The impacts are acts from design. The first half of the motion of each ball is under the direction of design. We mean by this the particular design of each player. But, at the instant of the collision of the balls upon each other, direction from design ceases, and the balls no longer obey the particular designs of the players, the ends or purposes intended by them are not accomplished, but frustrated, by *necessity*, or by the necessary action of the powers of inertia and elasticity, which are inherent in matter, and are not made by any design of a Creator for this special action, or to serve this special purpose, but would have existed in the materials of which the balls were made, although the players had never been born.

I have thus stated, by a simple example in physical action, what is meant by design and what by necessity; and that the latter may exist without any dependence upon the former. If I have given the statement with what may be thought, by some, unnecessary prolixity, I have only to say that I have found many minds to have a great difficulty in conceiving of necessity as acting altogether independent of design.

Let me now trace these principles as sources of action in Darwin's work or theory. Let us see how much there is of design acting to produce a foreseen end, and thus proving a reasoning and self-conscious Creator; and how much of mere blind power acting without rational design, or without a specific purpose or conscious foresight. Mr. Darwin has specified in a most clear and unmistakable manner the operation of his three great powers, or rather, the three great laws by which the organic power of life acts in the formation of an eye. (*See* p. 169.) Following the method he has pointed out, we will take a number of animals of the same species, in which the eye is not developed. They may have all the other senses, with the organs of nutrition, circulation, respiration, and locomotion. They all have a brain and nerves, and some of these nerves may be sensitive to light; but have no combination of retina, membranes, humors, etc., by which the distinct image of an object may be formed and conveyed by the optic nerve to the cognizance of the internal perception, or the mind. The animal in this case would be merely sensible of the difference between light and darkness. He would have no power of discriminating form, size, shape, or color, the difference of objects, and to gain from these a knowledge of their being useful or hurtful, friends or enemies. Up to this point there is no appearance of *necessity* upon the scene. The billiard-balls have not yet struck together, and we will suppose that none of the arguments that may be used to prove, from this organism, thus existing, that it could not have come into form and being without a creator acting to this end with intelligence and design, are opposed by anything that can be found in Darwin's theory; for, so far, Darwin's laws are supposed not to have come into operation. Give the animals, thus organized, food and room, and they may go on, from generation to generation, upon the same organic level. Those individuals that, from natural variation, are born with *light-nerves* a little more sensitive to light than their parents, will cross or interbreed with those who have the same organs a little less sensitive, and thus the mean standard will be kept up without any advancement. If our billiard-table were sufficiently extensive, i. e., infinite, the balls rolled from the corners would never meet, and the *necessity* which we have supposed to deflect them would never act.

The moment, however, that the want of space or food commences *natural selection* begins. Here the balls meet, and all future action is governed by *necessity.* The best forms, or those nerves most sensitive to light, connected with incipient membranes and humors for corneas and lenses, are picked out and preserved by natural selection, of necessity. All cannot live and propagate, and it is a necessity, obvious to all, that the weaker must perish, if the theory be true. Working on, in this way, through countless generations, the eye is at last formed in all its beauty and excellence. It must (always assuming that this theory is true) result from this combined action of natural variation, the struggle for life, and natural selection, with as much certainty as the balls, after collision, must pass to corners of the table different from those to which they were directed, and so far forth as the eye is formed by these laws, acting upward from the nerve merely sensitive to light, we can no more infer design, and from design a designer, than we can infer design in the direction of the billiard-balls after the collision. Both are sufficiently accounted for by blind powers acting under a blind necessity. Take away the struggle for life from the one, and the collision of the balls from the other—and neither of these was designed—and the animal would have gone on without eyes. The balls would have found the corners of the table to which they were first directed.

While, therefore, it seems to me clear that one who can find no proof of the existence of an intelligent Creator except through the evidence of design in the organic world, can find no evidence of such design in the construction of the eye, if it were constructed under the operation of Darwin's laws, I shall not for one moment contend that these laws are *incompatible* with design and a self-conscious, intelligent Creator. Such design might, indeed, have coexisted with the necessity or natural selection; and so the billiard-players might have designed the collision of their balls; but neither the formation of the eye, nor the path of the balls after collision, furnishes any sufficient proof of such design in either case.

One, indeed, who believes, from revelation or any other cause, in the existence of such a Creator, the fountain and source of all things in heaven above and in the earth beneath, will see in natural variation, the struggle for life, and natural selection, only the order or mode in which this Creator, in his own perfect wisdom, sees fit to act. Happy is he who can thus see and adore. But how many are there who have no such belief from intuition, or faith in revelation; but who have by careful and elaborate search in the physical, and more especially in the organic world, inferred, by induction, the existence of God from what has seemed to them the wonderful adaptation of the different organs and parts of the animal body to its, apparently, designed ends! Imagine a mind of this skeptical character, in all honesty and under its best reason, after finding itself obliged to reject the evidence of revelation, to commence a search after the Creator, in the light of natural theology. He goes through the proof for final cause and design, as given in a summary though clear, plain, and convincing form, in the pages of Paley and the "Bridgewater Treatises." The eye and the hand, those perfect instruments of optical and mechanical contrivance and adaptation, without the least waste or surplusage—these, say Paley and Bell, certainly prove a designing maker as much as the palace or the watch proves an architect or a watchmaker. Let this mind, in this state, cross Darwin's work, and find that, after a sensitive nerve or a rudimentary hoof or claw, no design is to be found. From this point upward the development is the mere necessary result of natural selection; and let him receive this law of natural selection as true, and where does he find himself? Before, he could refer the existence of the eye, for example, only to design, or chance. There was no other alternative. He rejected chance, as impossible. It must then be a design. But Darwin brings up another power, namely, natural selection, in place of this impossible chance. This not only may, but, according to Darwin, must of necessity produce an eye. It may indeed coexist with design, but it must exist and act and produce its results, even without design. Will such a mind, under such circumstances, infer the existence of the designer—God—when he can, at the same time, satisfactorily account for the thing produced, by the operation of this natural selection? It seems to me, therefore, perfectly evident that the substitution of natural selection, by necessity, for design in the formation of the organic world, is a step decidedly atheistical. It is in vain to say that Darwin takes the creation of organic life, in its simplest forms, to have been the work of the Deity. In giving up design in these highest and most complex forms of organization, which have always been relied upon as the crowning proof of the existence of an intelligent Creator, without whose intellectual power they could not have been brought into being, he takes a most decided step to banish a belief in the intelligent action of God from the organic world. The lower organisms will go next.

The atheist will say, Wait a little. Some future Darwin will show how the simple forms came *necessarily* from inorganic matter. This is but another step by which, according to Laplace, "the discoveries of science throw final causes further back."

A. G.—It is conceded that, if the two players in the supposed case were ignorant of each other's presence, the designs of both were frustrated, and from necessity. Thus far it is not needful to inquire whether this necessary consequence is an unconditional or a conditioned necessity, nor to require a more definite statement of the meaning attached to the word *necessity* as a supposed third alternative.

But, if the players knew of each other's presence, we could not infer from the result that the design of both or of either was frustrated. One of them may have intended to frustrate the other's design, and to effect his own. Or both may have been equally conversant with the properties of the matter and the relation of the forces concerned (whatever the cause, origin, or nature, of these forces and properties), and the result may have been according to the designs of both.

As you admit that they might or might not have designed the collision of their balls and its consequences, the question arises whether there is any way of ascertaining which of the two conceptions we may form about it is the true one. Now, let it be remarked that *design* can never be *demonstrated.* Witnessing the act does not make known the *design*, as we have seen in the case assumed for the basis of the argument. The word of the actor is not proof; and that source of evidence is excluded from the cases in question. The only way left, and the only possible way in cases where testimony is out of the question, is to infer the design from the result, or from arrangements which strike us as *adapted* or *intended* to produce a certain result, which affords a presumption of design. The strength of this presumption may be zero, or an even chance, as perhaps it is in the assumed case; but the probability of design will increase with the particularity of the act, the specialty of the arrangement or machinery, and with the number of identical or yet more of similar and analogous instances, until it rises to a moral certainty—i. e., to a conviction which practically we are as unable to resist as we are to deny the cogency of a mathematical demonstration. A single instance, or set of instances, of a comparatively simple arrangement might suffice. For instance, we should not doubt that a pump was designed to raise water by the moving of the handle. Of course, the conviction is the stronger, or at least the sooner arrived at, where we can imitate the arrangement, and ourselves produce the result at will, as we could with a pump, and also with the billiard-balls.

And here I would suggest that your billiard-table, with the case of collision, answers well to a machine. In both a result is produced by indirection—by applying a force out of line of the ultimate direction. And, as I should feel as confident that a man intended to raise water who was working a pump-handle, as if he were bringing it up in pailfuls from below by means of a ladder, so, after due examination of the billiard-table and its appurtenances, I should probably think it likely that the effect of the rebound was expected and intended no less than that of the immediate impulse. And a similar inspection of arrangements and results in Nature would raise at least an equal presumption of design.

You allow that the rebound might have been intended, but you require proof that it was. We agree that a single such instance affords no evidence either way. But how would it be if you saw the men doing the same thing over and over? and if they varied it by other arrangements of the balls or of the blow, and these were followed by analogous results? How if you at length discovered a profitable end of the operation, say the winning of a wager? So in the counterpart case of natural selection: must we not infer intention from the arrangements and the results? But I will take another case of the very same sort, though simpler, and better adapted to illustrate natural selection; because the change of direction—your necessity—acts gradually or successively, instead of abruptly.

Suppose I hit a man standing obliquely in my rear, by throwing forward a crooked stick, called a boomerang. How could he know whether the blow was intentional or not? But suppose I had been known to throw boomerangs before; suppose that, on different occasions, I had before wounded persons by the same, or other indirect and apparently aimless actions; and suppose that an object appeared to be gained in the result—that definite ends were attained—would it not at length be inferred that my assault, though indirect, or apparently indirect, was designed?

To make the case more nearly parallel with those it is brought to illustrate, you have only to suppose that, although the boomerang thrown by me went forward to a definite place, and at least appeared to subserve a purpose, and the bystanders, after a while, could get traces of the mode or the empirical law of its flight, yet they could not themselves do anything with it. It was quite beyond their power to use it. Would they doubt, or deny *my* intention, on that account? No: they would insist that design on my part must be presumed from the nature of the results; that, though design *may* have been wanting in any one case, yet the repetition of the result, and from different positions and under varied circumstances, showed that there *must* have been design.

Moreover, in the way your case is stated, it seems to concede the most important half of the question, and so affords a presumption for the rest, on the side of design. For you seem to assume an actor, a designer, accomplishing his design in the first instance. You —a bystander—infer that the player effected his design in sending the first ball to the pocket before him. You infer this from observation alone. Must you not from a continuance of the same observation equally infer a common design of the two players in the complex result, or a design of one of them to frustrate the design of the other? If you grant a designing actor, the presumption of design is as strong, or upon continued observation of instances soon becomes as strong, in regard to the deflection of the balls, or variation of the species, as it was for the result of the first impulse or for the production of the original animal, etc.

But, in the case to be illustrated, we do not see the player. We see only the movement of the balls. Now, if the contrivances and adaptations referred to (p. 229) really do "prove a designer as much as the palace or the watch proves an architect or a watchmaker"—as Paley and Bell argue, and as your skeptic admits, while the alternative is between design and chance—then they prove it with all the proof the case is susceptible of, and with complete conviction. For we cannot doubt that the watch had a watchmaker. And if they prove it on the supposition that the unseen operator acted *immediately*—i. e., that the player directly impelled the balls in the directions we see them moving, I insist that this proof is not impaired by our ascertaining that he acted *mediately*—i. e., that the present state or form of the plants or animals, like the present position of the billiard-balls, resulted from the collision of the individuals with one another, or with the surroundings. The original impulse, which we once supposed was in the line of the observed movement, only proves to have been in a different direction; but the series of movements took place with a series of results, each and all of them none the less determined, none the less designed.

Wherefore, when, at the close, you quote Laplace, that "the discoveries of science throw final causes farther back," the most you can mean is, that they constrain us to look farther back for the impulse. They do not at all throw *the argument for design* farther back, in the sense of furnishing evidence or presumption that only the primary impulse was designed, and that all the rest followed from chance or necessity.

Evidence of design, I think you will allow, everywhere is drawn from the observation of adaptations and of results, and has really nothing to do with anything else, except where you can take the *word* for the *will*. And in that case you have not *argument for design*, but *testimony*. In Nature we have no testimony; but the argument is overwhelming.

Now, note that the argument of the olden time—that of Paley, etc., which your skeptic found so convincing—was always the argument for design in the movement of the balls *after deflection*. For it was drawn from animals produced by generation, not by creation, and through a long succession of generations or deflections. Wherefore, if the argument for design is perfect in the case of an animal derived from a long succession of individuals as nearly alike as offspring is generally like parents and grandparents, and if this argument is not weakened when a variation, or series or variations, has occurred in the course, as great as any variations we know of among domestic cattle, how then is it weakened by the supposition, or by the likelihood, that the variations have been twice or thrice as great as we formerly supposed, or because the variations have been "picked out," and a few of them preserved as breeders of still other variations, by natural selection?.

Finally let it be noted that your element of *necessity* has to do, so far as we know, only with the picking out and preserving of certain changing forms, i. e., with the natural selection. This selection, you may say, must happen under the circumstances. This is a necessary result of the collision of the balls; and these results can be predicted. If the balls strike so and so, they will be deflected so and so. But the *variation* itself is of the nature of an origination. It answers well to the original impulse of the balls, or to a series of such impulses. We cannot predict what particular new variation will occur from any observation of the past. Just as the first impulse was given to the balls at a point out of sight, so the inpulse which resulted in the variety or new form was given at a point beyond observation, and is equally mysterious or unaccountable, except on the supposition of an ordaining will. The parent had not the peculiarity of the variety, the progeny has. Between the two is the dim or obscure region of the formation of a new individual, in some unknown part of which, and in some wholly unknown way, the difference is intercalated. To introduce necessity here is gratuitous and unscientific; but here you must have it to make your argument valid.

I agree that, judging from the past, it is not improbable that variation itself may be hereafter shown to result from physical causes. When it is so shown, you may extend your necessity into this region, but not till then. But the whole course of scientific discovery goes to assure us that the discovery of the cause of variation will be only a resolution of variation into two factors: one, the immediate secondary cause of the changes, which so far explains them; the other an unresolved or unexplained phenomenon, which will then stand just where the product, variation, stands now, only that it will be one step nearer to the efficient cause.

This line of argument appears to me so convincing, that I am bound to suppose that it does not meet your case. Although you introduced players to illustrate what design is, it is probable that you did not intend, and would not accept, the parallel which your supposed case suggested. When you declare that the proof of design in the eye and the hand, as given by Paley and Bell, was convincing, you mean, of course, that it was convincing, so long as the question was between *design* and *chance*, but that now another alternative is offered, one which obviates the force of those arguments, and may account for the actual results without design. I do not clearly apprehend this third alternative.

Will you be so good, then, as to state the grounds upon which you conclude that the supposed proof of design from the eye, or the hand, as it stood before Darwin's theory was promulgated, would be invalidated by the admission of this new theory?

D. T.—As I have ever found you, in controversy, meeting the array of your opponent fairly and directly, without any attempt to strike the body of his argument through an unguarded joint in the phraseology, I was somewhat surprised at the course taken in your answer to my statement on Darwin's theory. You there seem to suppose that I instanced the action of the billiard balls and players as a parallel, throughout, to the formation of the organic world. Had it occurred to me that such an application might be supposed to follow legitimately from my introduction of this action, I should certainly have stated that I did not intend, and should by no means accede to, that construction. My purpose in bringing the billiard-table upon the scene was to illustrate, by example, *design* and *necessity*, as different and independent sources from which results, it might indeed be identical results, may be derived. All the conclusions, therefore, that you have arrived at through this misconception or misapplication of my illustration, I cannot take as an answer to the matter stated or intended to be stated by me. Again, following this misconception, you suppose the skeptic (instanced by me as revealing through the evidence of design, exhibited in the structure of the eye, for its designer, God) as bringing to the examination a belief in the existence of design in the construction of the animals as they existed up to the moment when the eye was, according to my supposition, added to the heart, stomach, brain, etc. By skeptic I, of course, intended one who doubted the existence of design in every organic structure, or at least required proof of such design. Now, as the watch may be instanced as a more complete exhibition of design than a flint knife or an hour-glass, I selected, after the example of Paley, the eye, as exhibiting by its complex but harmonious arrangements a higher evidence of design and a designer than is to be found in a nerve sensitive to light, or any mere rudimentary part or organ. I could not mean by skeptic one who believed in design so far as a claw, or a nerve sensitive to light, was concerned, but doubted all above. For one who believes in design at all will not fail to recognize it in a hand or an eye. But I need not extend these remarks, as you acknowledge in the sequel to your argument that you may not have suited it to the case as I had stated it.

You now request me to "state the grounds upon which I conclude that the supposed proof of design from the eye and the hand, as it stood before Darwin's theory was promulgated, is invalidated by the admission of that theory." It seems to me that a sufficient answer to this question has already been made in the last part of my former paper; but, as you request it, I will go over the leading points as there given, with more minuteness of detail.

Let us, then, suppose a skeptic, one who is yet considering and doubting of the existence of God, having already concluded that the testimony from any and all revelation is insufficient, and having rejeeted what is called the *a priori* arguments brought forward in natural theology, and pertinaciously insisted upon by Dr. Clark and others, turning as a last resource to the argument from design in the organic world. Voltaire tells him that a palace could not exist without an architect to design it. Dr. Paley tells him that a watch proves the design of a watchmaker. He thinks this very reasonable, and, although he sees a difference between the works of Nature and those of mere human art, yet if he can find in any organic body, or part of a body, the same adaptation to its use that he finds in a watch, this truth will go very far toward proving, if it is not entirely conclusive, that, in making it, the powers of life by which it grew were directed by an intelligent, reasoning master. Under the guidance of Paley he takes an eye, which, although an optical, and not a mechanical instrument like the watch, is as well adapted to testify to design. He sees, first, that the eye is transparent when every other part of the body is opaque. Was this the result of a mere Epicurean or Lucretian "fortuitous concourse" of living "atoms?" He is not yet certain it might not be so. Next he sees that it is spherical, and that this convex form alone is capable of changing the direction of the light which proceeds from a distant body, and of collecting it so as to form a distinct image within its globe. Next he sees at the exact place where this image must be formed a curtain of nerve-work, ready to receive and convey it, or excite from it, in its own mysterious way, an idea of it in the mind. Last of all, he comes to the crystalline lens. Now, he has before learned that without this lens an eye would by the aqueous and vitreous humors alone form an image upon the retina, but this image would be indistinct from the light not being sufficiently refracted, and likewise from having a colored fringe round its edges. This last effect is attributable to the refrangibility of light, that is, to some of the colors being more refracted than others. He likewise knows that more then a hundred years ago Mr. Dollond having found out, after many experiments, that some kinds of glass have the power of dispersing light, for each degree of its refraction, much more than other kinds, and that on the discovery of this fact he contrived to make telescopes in which he passed the light through two object-glasses successively, one of which he made of crown and one of flint glass, so ground and adapted to each other that the greater dispersion produced by the substance of one should be corrected by the smaller dispersion of the other. This contrivance corrected entirely the colored images which had rendered all previous telescopes very imperfect. He finds in this invention all the elements of design, as it appeared in the thought and action of a human designer. First, conjecture of certain laws or facts in optics. Then, experiment proving these laws or facts. Then, the contrivance and formation of an instrument by which those laws or facts must produce a certain sought result.

Thus enlightened, our skeptic turns to his crystalline lens to see if he can discover the work of a Dollond in this. Here he finds that an eye, having a crystalline lens placed between the humors, not only refracts the light more than it would be refracted by the humors alone, but that, in this combination of humors and lens, the colors are as completely corrected as in the combination of Dollond's telescope. Can it be that there was no design, no designer, directing the powers of life in the formation of this wonderful organ? Our skeptic is aware that, in the arts of man, great aid has been, sometimes, given by chance, that is, by the artist or workman observing some fortuitous combination, form, or action, around him. He has heard it said that the chance arrangement of two pairs of spectacles, in the shop of a Dutch optician, gave the direction for constructing the first telescope. Possibly, in time, say a few geological ages, it might in some optician's shop have brought about a combination of flint and crown glass which, together, should have been achromatic. But the space between the humors of the eye is not an optician's shop where object-glasses of all kinds, shapes, and sizes, are placed by chance, in all manner of relations and positions. On the hypothesis under which our skeptic is making his examination—the eye having been completed in all but the formation of the lens—the place which the lens occupies when completed was filled with parts of the humors and plane membrane, homogeneous in texture and surface, presenting, therefore, neither the variety of the materials nor forms which are contained in the optician's shop for chance to make its combinations with. How, then, could it be cast of a combination not before used, and fashioned to a shape different from that before known, and placed in exact combination with all the parts before enumerated, with many others not even mentioned? He sees no parallelism of condition, then, by which chance could act in forming a crystalline lens, which answers to the condition of an optician's shop, where it might be possible in many ages for chance to combine existing forms into an achromatic object-glass.

Considering, therefore, the eye thus completed and placed in its bony case and provided with its muscles, its lids, its tear-ducts, and all its other elaborate and curious appendages, and, a thousand times more wonderful still, without being encumbered with a single superfluous or useless part, can he say that this could be the work of chance? The improbability of this is so great, and consequently the evidence of design is so strong, that he is about to seal his verdict in favor of design, when he opens Mr. Darwin's book.

There he finds that an eye is no more than a vital aggregation or growth, directed, not by design nor chance, but moulded by natural variation and natural selection, through which it must, necessarily, have been developed and formed. Particles or atoms being aggregated by the blind powers of life, must become under the given conditions, by natural variation and natural selection, eyes, without design, as certainly as the red billiard-ball went to the west pocket, by the powers of inertia and elasticity, without the design of the hand that put it in motion. (*See* Darwin, p. 169.)

Let us lay before our skeptic the way in which we may suppose that Darwin would trace the operation of life, or the vital force conforming to these laws. In doing this we need not go through with the formation of the several membranes, humors, etc., but take the crystalline lens as the most curious and nicely arranged and adapted of all the parts, and as giving, moreover, a close parallel, in the end produced, to that produced by design, by a human designer, Dollond, in forming his achromatic object-glass. If it can be shown that natural variation and natural selection were capable of forming the crystalline lens, it will not be denied that they were capable of forming the iris, the sclerotica, the aqueous humors, or any and all the other parts. Suppose, then, that we have a number of animals, with eyes yet wanting the crystalline. In this state the animals can see, but dimly and imperfectly, as a man sees after having been couched. Some of the offspring of these animals have, by *natural variation*, merely a portion of the membrane which separates the aqueous from the vitreous humor a little thickened in its middle part, a little swelled out. This refracts the light a little more than it would be refracted by a membrane in which no such swelling existed, and not only so, but, in combination with the humors, it corrects the errors of dispersion and makes the image somewhat more colorless. All the young animals that have this swelled membrane see more distinctly than their parents or brethren. They, therefore, have an advantage over them in the *struggle for life*. They can obtain food more easily; can find their prey, and escape from their enemies with greater facility than their kindred. This thickening and rounding of the membrane goes on from generation to generation by natural variation; natural selection all the while "picking out with unerring skill all the improvements, through countless generations," until at length it is found that the membrane has become a perfect crystalline lens. Now, where is the design in all this? The membrane was not thickened and rounded to the end that the image should be more distinct and colorless; but, being thickened and rounded by the operation of natural variation, *inherent* in generation, natural selection *of necessity* produced the result that we have seen. The same result was thus produced *of necessity*, in the eye, that Dollond came at, in the telescope, with design, through painful guessing, reasoning, experimenting, and forming.

Suppose our skeptic to believe in all this power of natural selection; will he now seal up his verdict for design, with the same confidence that he would before he heard of Darwin? If not, then "the supposed proof from design is invalidated by Darwin's theory."

A. G.—Waiving incidental points and looking only to the gist of the question, I remark that the argument for design as against chance, in the formation of the eye, is most convincingly stated in your argument. Upon this and upon numerous similar arguments the whole question we are discussing turns. So, if the skeptic was about to seal his verdict in favor of design, and a designer, when Darwin's book appeared, why should his verdict now be changed or withheld? All the facts about the eye, which convinced him that the organ was designed, remain just as they were. His conviction was not produced through testimony or eyewitness, but design was irresistibly inferred from the evidence of contrivance in the eye itself.

Now, if the eye as it is, or has become, so convincingly argued design, why not each particular step or part of this result? If the production of a perfect crystalline lens in the eye—you know not how—as much indicated design as did the production of a Dollond achromatic lens—you understand how—then why does not "the swelling out" of a particular portion of the membrane behind the iris—caused you know not how—which, by "correcting the errors of dispersion and making the image somewhat more colorless," enabled the "young animals to see more distinctly than their parents or brethren," equally indicate design —if not as much as a perfect crystalline, or a Dollond compound lens, yet as much as a common spectacle-glass?

Darwin only assures you that what you may have thought was done directly and at once was done indirectly and successively. But you freely admit that indirection and succession do not invalidate design, and also that Paley and all the natural theologians drew the arguments which convinced your skeptic wholly from eyes indirectly or naturally produced.

Recall a woman of a past generation and show her a web of cloth; ask her how it was made, and she will say that the wool or cotton was carded, spun, and woven by hand. When you tell her it was not made by manual labor, that probably no hand has touched the materials throughout the process, it is possible that she might at first regard your statement as tantamount to the assertion that the cloth was made without design. If she did, she would not credit your statement. If you patiently explained to her the theory of carding-machines, spinning-jennies, and power-looms, would her reception of your explanation weaken her conviction that the cloth was the result of design? It is certain that she would believe in design as firmly as before, and that this belief would be attended by a higher conception and reverent admiration of a wisdom, skill, and power greatly beyond anything she had previously conceived possible.

Wherefore, we may insist that, for all that yet

appears, the argument for design, as presented by the natural theologians, is just as good now, if we accept Darwin's theory, as it was before that theory was promulgated; and that the skeptical juryman, who was about to join the other eleven in a unanimous verdict in favor of design, finds no good excuse for keeping the court longer waiting.[1]

[[1] To parry an adversary's thrust at a vulnerable part, or to show that it need not be fatal, is an incomplete defense. If the discussion had gone on, it might, perhaps, have been made to appear that the Darwinian hypothesis, so far from involving the idea of necessity (except in the sense that everything is of necessity), was based upon the opposite idea, that of contingency.]

III.

NATURAL SELECTION NOT INCONSISTENT WITH NATURAL THEOLOGY.

ATLANTIC MONTHLY FOR *July*, *August*, AND *October*, 1860, REPRINTED IN 1861.

I.

NOVELTIES are enticing to most people; to us they are simply annoying. We cling to a long-accepted theory, just as we cling to an old suit of clothes. A new theory, like a new pair of breeches (the *Atlantic* still affects the older type of nether garment), is sure to have hard-fitting places; or, even when no particular fault can be found with the article, it oppresses with a sense of general discomfort. New notions and new styles worry us, till we get well used to them, which is only by slow degrees.

Wherefore, in Galileo's time, we might have helped to proscribe, or to burn—had he been stubborn enough to warrant cremation—even the great pioneer of inductive research; although, when we had fairly recovered our composure, and had leisurely excogitated the matter, we might have come to conclude that the new doctrine was better than the old one, after all, at least for those who had nothing to unlearn.

Such being our habitual state of mind, it may well

be believed that the perusal of the new book "On the Origin of Species by Means of Natural Selection" left an uncomfortable impression, in spite of its plausible and winning ways. We were not wholly unprepared for it, as many of our contemporaries seem to have been. The scientific reading in which we indulge as a relaxation from severer studies had raised dim forebodings. Investigations about the succession of species in time, and their actual geographical distribution over the earth's surface, were leading up from all sides and in various ways to the question of their origin. Now and then we encountered a sentence, like Prof. Owen's "axiom of the continuous operation of the ordained becoming of living things," which haunted us like an apparition. For, dim as our conception must needs be as to what such oracular and grandiloquent phrases might really mean, we felt confident that they presaged no good to old beliefs. Foreseeing, yet deprecating, the coming time of trouble, we still hoped that, with some repairs and makeshifts, the old views might last out our days. *Après nous le déluge.* Still, not to lag behind the rest of the world, we read the book in which the new theory is promulgated. We took it up, like our neighbors, and, as was natural, in a somewhat captious frame of mind.

Well, we found no cause of quarrel with the first chapter. Here the author takes us directly to the barn-yard and the kitchen-garden. Like an honorable rural member of our General Court, who sat silent until, near the close of a long session, a bill requiring all swine at large to wear pokes was introduced, when he claimed the privilege of addressing the house, on the proper ground that he had been "brought up among the pigs, and knew all about them"—so we were brought up among cows and cabbages; and the lowing of cattle, the cackle of hens, and the cooing of pigeons, were sounds native and pleasant to our ears. So "Variation under Domestication" dealt with familiar subjects in a natural way, and gently introduced "Variation under Nature," which seemed likely enough. Then follows "Struggle for Existence"—a principle which we experimentally know to be true and cogent—bringing the comfortable assurance, that man, even upon Leviathan Hobbes's theory of society, is no worse than the rest of creation, since all Nature is at war, one species with another, and the nearer kindred the more internecine—bringing in thousand-fold confirmation and extension of the Malthusian doctrine that population tends far to outrun means of subsistence throughout the animal and vegetable world, and has to be kept down by sharp preventive checks; so that not more than one of a hundred or a thousand of the individuals whose existence is so wonderfully and so sedulously provided for ever comes to anything, under ordinary circumstances; so the lucky and the strong must prevail, and the weaker and ill-favored must perish; and then follows, as naturally as one sheep follows another, the chapter on "Natural Selection," Darwin's *cheval de bataille*, which is very much the Napoleonic doctrine that Providence favors the strongest battalions—that, since many more individuals are born than can possibly survive, those individuals and those variations which possess any advantage, however slight, over the rest, are in the long-run sure to survive, to propagate, and to occupy the limited field, to the exclusion or destruction of the weaker brethren. All this we pondered, and could not much object to. In fact, we began to contract a liking for a system which at the outset illustrates the advantages of good breeding, and which makes the most "of every creature's best."

Could we "let by-gones be by-gones," and, beginning now, go on improving and diversifying for the future by natural selection, could we even take up the theory at the introduction of the actually existing species, we should be well content; and so, perhaps, would most naturalists be. It is by no means difficult to believe that varieties are incipient or possible species, when we see what trouble naturalists, especially botanists, have to distinguish between them—one regarding as a true species what another regards as a variety; when the progress of knowledge continually increases, rather than diminishes, the number of doubtful instances; and when there is less agreement than ever among naturalists as to what is the basis in Nature upon which our idea of species reposes, or how the word is to be defined. Indeed, when we consider the endless disputes of naturalists and ethnologists over the human races, as to whether they belong to one species or to more, and, if to more, whether to three, or five, or fifty, we can hardly help fancying that both may be right—or rather, that the uni-humanitarians would have been right many thousand years ago, and the multi-humanitarians will be several thousand years later; while at present the safe thing to say is, that probably there is some truth on both sides.

"Natural selection," Darwin remarks, "leads to divergence of character; for the more living beings can be supported on the same area, the more they diverge in structure, habits, and constitution" (a principle which, by-the-way, is paralleled and illustrated by the diversification of human labor); and also leads to much extinction of intermediate or unimproved forms. Now, though this divergence may "steadily tend to increase," yet this is evidently a slow process in Nature, and liable to much counteraction wherever man does not interpose, and so not likely to work much harm for the future. And if natural selection, with artificial to help it, will produce better animals and better men than the present, and fit them better "to the conditions of existence," why, let it work, say we, to the top of its bent. There is still room enough for improvement. Only let us hope that it always works for good: if not, the divergent lines on Darwin's lithographic diagram of "Transmutation made Easy," ominously show what small deviations from the straight path may come to in the end.

The prospect of the future, accordingly, is on the whole pleasant and encouraging. It is only the backward glance, the gaze up the long vista of the past, that reveals anything alarming. Here the lines converge as they recede into the geological ages, and point to conclusions which, upon the theory, are inevitable, but hardly welcome. The very first step backward makes the negro and the Hottentot our blood-relations—not that reason or Scripture objects to that, though pride may. The next suggests a closer association of our ancestors of the olden time with "our poor relations" of the quadrumanous family than we like to acknowledge. Fortunately, however—even if we must account for him scientifically—man with his two feet stands upon a foundation of his own. Intermediate links between the *Bimana* and the *Quadrumana* are lacking altogether; so that, put the genealogy of the brutes upon what footing you will, the four-handed races will not serve for our forerunners—at least, not until some monkey, live or fossil, is producible with great-toes, instead of thumbs, upon his nether extremities; or until some lucky geologist turns up the bones of his ancestor and prototype in France or England, who was so busy "napping the chuckie-stanes" and chipping out flint knives and arrow-heads in the time of the drift, very many ages ago—before the British Channel existed, says Lyell[1]—and until these men of the olden time are shown to have worn their great-toes in the divergent and thumb-like fashion. That would be evidence indeed: but, until some testimony of the sort is produced, we must needs believe in the separate and special creation of man, however it may have been with the lower animals and with plants.

No doubt, the full development and symmetry of Darwin's hypothesis strongly suggest the evolution of

[1] *Vide* "Proceedings of the British Association for the Advancement of Science," 1859, and London *Athenæum*, *passim.* It appears to be conceded that these "celts" or stone knives are artificial productions, and apparently of the age of the mammoth, the fossil rhinoceros, etc.

the human no less than the lower animal races out of some simple primordial animal—that all are equally "lineal descendants of some few beings which lived long before the first bed of the Silurian system was deposited." But, as the author speaks disrespectfully of spontaneous generation, and accepts a supernatural beginning of life on earth, in some form or forms of being which included potentially all that have since existed and are yet to be, he is thereby not warranted to extend his inferences beyond the evidence or the fair probability. There seems as great likelihood that one special origination should be followed by another upon fitting occasion (such as the introduction of man), as that one form should be transmuted into another upon fitting occasion, as, for instance, in the succession of species which differ from each other only in some details. To compare small things with great in a homely illustration: man alters from time to time his instruments or machines, as new circumstances or conditions may require and his wit suggest. Minor alterations and improvements he adds to the machine he possesses; he adapts a new rig or a new rudder to an old boat: this answers to *Variation.* "Like begets like," being the great rule in Nature, if boats could engender, the variations would doubtless be propagated, like those of domestic cattle. In course of time the old ones would be worn out or wrecked; the best sorts would be chosen for each particular use, and further improved upon; and so the primordial boat be developed into the scow, the skiff, the sloop, and other species of water-craft—the very diversification, as well as the successive improvements, entailing the disappearance of intermediate forms, less adapted to any one particular purpose; wherefore these go slowly out of use, and become extinct species: this is *Natural Selection*. Now, let a great and important advance be made, like that of steam navigation: here, though the engine might be added to the old vessel, yet the wiser and therefore the actual way is to make a new vessel on a modified plan: this may answer to *Specific Creation*. Anyhow, the one does not necessarily exclude the other. Variation and natural selection may play their part, and so may specific creation also. Why not?

This leads us to ask for the reasons which call for this new theory of transmutation. The beginning of things must needs lie in obscurity, beyond the bounds of proof, though within those of conjecture or of analogical inference. Why not hold fast to the customary view, that all species were directly, instead of indirectly, created after their respective kinds, as we now behold them—and that in a manner which, passing our comprehension, we intuitively refer to the supernatural? Why this continual striving after "the unattained and dim?" why these anxious endeavors, especially of late years, by naturalists and philosophers of various schools and different tendencies, to penetrate what one of them calls "that mystery of mysteries," the origin of species?

To this, in general, sufficient answer may be found in the activity of the human intellect, "the delirious yet divine desire to know," stimulated as it has been by its own success in unveiling the laws and processes of inorganic Nature; in the fact that the principal triumphs of our age in physical science have consisted in tracing connections where none were known before, in reducing heterogeneous phenomena to a common cause or origin, in a manner quite analogous to that of the reduction of supposed independently originated species to a common ultimate origin—thus, and in various other ways, largely and legitimately extending the domain of secondary causes. Surely the scientific mind of an age which contemplates the solar system as evolved from a common revolving fluid mass—which, through experimental research, has come to regard light, heat, electricity, magnetism, chemical affinity, and mechanical power as varieties or derivative and convertible forms of one force, instead of independent species—which has brought the so-called elementary kinds of matter, such as the metals, into kindred groups, and pertinently raised the question, whether the members of each group may not be mere varieties of one species—and which speculates steadily in the direction of the ultimate unity of matter, of a sort of prototype or simple element which may be to the ordinary species of matter what the *Protozoa* or what the component cells of an organism are to the higher sorts of animals and plants—the mind of such an age cannot be expected to let the old belief about species pass unquestioned. It will raise the question, how the diverse sorts of plants and animals came to be as they are and where they are, and will allow that the whole inquiry transcends its powers only when all endeavors have failed. Granting the origin to be supernatural, or miraculous even, will not arrest the inquiry. All real origination, the philosophers will say, is supernatural; their very question is, whether we have yet gone back to the origin, and can affirm that the present forms of plants and animals are the primordial, the miraculously created ones. And, even if they admit that, they will still inquire into the order of the phenomena, into the form of the miracle. You might as well expect the child to grow up content with what it is told about the advent of its infant brother. Indeed, to learn that the new-comer is the gift of God, far from lulling inquiry, only stimulates speculation as to how the precious gift was bestowed. That questioning child is father to the man—is philosopher in short-clothes.

Since, then, questions about the origin of species will be raised, and have been raised—and since the theorizings, however different in particulars, all proceed upon the notion that one species of plant or animal is somehow derived from another, that the different sorts which now flourish are lineal (or unlineal) descendants of other and earlier sorts—it now concerns us to ask, What are the grounds in Nature, the admitted facts, which suggest hypotheses of derivation in some shape or other? Reasons there must be, and plausible ones, for the persistent recurrence of theories upon this genetic basis. A study of Darwin's book, and a general glance at the present state of the natural sciences, enable us to gather the following as among the most suggestive and influential. We can only enumerate them here, without much indication of their particular bearing. There is—

1. The general fact of variability, and the general tendency of the variety to propagate its like—the

patent facts that all species vary more or less; that domesticated plants and animals, being in conditions favorable to the production and preservation of varieties, are apt to vary widely; and that, by interbreeding, any variety may be fixed into a race, that is, into a variety which comes true from seed. Many such races, it is allowed, differ from each other in structure and appearance as widely as do many admitted species; and it is practically very difficult, even impossible, to draw a clear line between races and species. Witness the human races, for instance. Wild species also vary, perhaps about as widely as those of domestication, though in different ways. Some of them apparently vary little, others moderately, others immoderately, to the great bewilderment of systematic botanists and zoölogists, and increasing disagreement as to whether various forms shall be held to be original species or strong varieties. Moreover, the degree to which the descendants of the same stock, varying in different directions, may at length diverge, is unknown. All we know is, that varieties are themselves variable, and that very diverse forms have been educed from one stock.

2. Species of the same genus are not distinguished from each other by equal amounts of difference. There is diversity in this respect analogous to that of the varieties of a polymorphous species, some of them slight, others extreme. And in large genera the unequal resemblance shows itself in the clustering of the species around several types or central species, like satellites around their respective planets. Obviously suggestive this of the hypothesis that they were satellites, not thrown off by revolution, like the

moons of Jupiter, Saturn, and our own solitary moon, but gradually and peacefully detached by divergent variation. That such closely-related species may be only varieties of higher grade, earlier origin, or more favored evolution, is not a very violent supposition. Anyhow, it was a supposition sure to be made.

3. The actual geographical distribution of species upon the earth's surface tends to suggest the same notion. For, as a general thing, all or most of the species of a peculiar genus or other type are grouped in the same country, or occupy continuous, proximate, or accessible areas. So well does this rule hold, so general is the implication that kindred species are or were associated geographically, that most trustworthy naturalists, quite free from hypotheses of transmutation, are constantly inferring former geographical continuity between parts of the world now widely disjoined, in order to account thereby for certain generic similarities among their inhabitants; just as philologists infer former connection of races, and a parent language, to account for generic similarities among existing languages. Yet no scientific explanation has been offered to account for the geographical association of kindred species, except the hypothesis of a common origin.

4. Here the fact of the antiquity of creation, and in particular of the present kinds of the earth's inhabitants, or of a large part of them, comes in to rebut the objection that there has not been time enough for any marked diversification of living things through divergent variation—not time enough for varieties to have diverged into what we call species.

So long as the existing species of plants and animals were thought to have originated a few thousand years ago, and without predecessors, there was no room for a theory of derivation of one sort from another, nor time enough even to account for the establishment of the races which are generally believed to have diverged from a common stock. Not so much that five or six thousand years was a short allowance for this; but because some of our familiar domesticated varieties of grain, of fowls, and of other animals, were pictured and mummified by the old Egyptians more than half that number of years ago, if not earlier. Indeed, perhaps the strongest argument for the original plurality of human species was drawn from the identification of some of the present races of men upon these early historical monuments and records.

But this very extension of the current chronology, if we may rely upon the archæologists, removes the difficulty by opening up a longer vista. So does the discovery in Europe of remains and implements of prehistoric races of men, to whom the use of metals was unknown—men of the *stone age*, as the Scandinavian archæologists designate them. And now, "axes and knives of flint, evidently wrought by human skill, are found in beds of the drift at Amiens (also in other places, both in France and England), associated with the bones of extinct species of animals." These implements, indeed, were noticed twenty years ago; at a place in Suffolk they have been exhumed from time to time for more than a century; but the full confirmation, the recognition of the age of the deposit in which the implements occur, their abundance, and the appreciation of their bearings upon most interesting questions, belong to the present time. To complete the connection of these primitive people with the fossil ages, the French geologists, we are told, have now "found these axes in Picardy associated with remains of *Elephas primigenius*, *Rhinoceros tichorhinus*, *Equus fossilis*, and an extinct species of *Bos*."[1] In plain language, these workers in flint lived in the time of the mammoth, of a rhinoceros now extinct, and along with horses and cattle unlike any now existing —specifically different, as naturalists say, from those with which man is now associated. Their connection with existing human races may perhaps be traced through the intervening people of the stone age, who were succeeded by the people of the bronze age, and these by workers in iron.[2] Now, various evidence carries back the existence of many of the present lower species of animals, and probably of a larger number of plants, to the same drift period. All agree that this was very many thousand years ago. Agassiz tells us that the same species of polyps which are now building coral walls around the present peninsula of Florida actually made that peninsula, and have been building there for many thousand centuries.

5. The overlapping of existing and extinct species, and the seemingly gradual transition of the life of the drift period into that of the present, may be turned to

[1] *See* "Correspondence of M. Nicklès," in *American Journal of Science and Arts*, for March, 1860.

[2] *See* Morlot, "Some General Views on Archæology," in *American Journal of Science and Arts*, for January, 1860, translated from "Bulletin de la Société Vaudoise," 1859.

the same account. Mammoths, mastodons, and Irish elks, now extinct, must have lived down to human, if not almost to historic times. Perhaps the last dodo did not long outlive his huge New Zealand kindred. The auroch, once the companion of mammoths, still survives, but owes his present and precarious existence to man's care. Now, nothing that we know of forbids the hypothesis that some new species have been independently and supernaturally created within the period which other species have survived. Some may even believe that man was created in the days of the mammoth, became extinct, and was recreated at a later date. But why not say the same of the auroch, contemporary both of the old man and of the new? Still it is more natural, if not inevitable, to infer that, if the aurochs of that olden time were the ancestors of the aurochs of the Lithuanian forests, so likewise were the men of that age the ancestors of the present human races. Then, whoever concludes that these primitive makers of rude flint axes and knives were the ancestors of the better workmen of the succeeding stone age, and these again of the succeeding artificers in brass and iron, will also be likely to suppose that the *Equus* and *Bos* of that time, different though they be, were the remote progenitors of our own horses and cattle. In all candor we must at least concede that such considerations suggest a genetic descent from the drift period down to the present, and allow time enough—if time is of any account—for variation and natural selection to work out some appreciable results in the way of divergence into races, or even into so-called species. Whatever might have been thought, when geological time was supposed to be separated from the present era by a clear line, it is now certain that a gradual replacement of old forms by new ones is strongly suggestive of some mode of origination which may still be operative. When species, like individuals, were found to die out one by one, and apparently to come in one by one, a theory for what Owen sonorously calls "the continuous operation of the ordained becoming of living things" could not be far off.

That all such theories should take the form of a derivation of the new from the old seems to be inevitable, perhaps from our inability to conceive of any other line of secondary causes in this connection. Owen himself is apparently in travail with some transmutation theory of his own conceiving, which may yet see the light, although Darwin's came first to the birth. Different as the two theories will probably be, they cannot fail to exhibit that fundamental resemblance in this respect which betokens a community of origin, a common foundation on the general facts and the obvious suggestions of modern science. Indeed—to turn the point of a pungent simile directed against Darwin—the difference between the Darwinian and the Owenian hypotheses may, after all, be only that between homœopathic and heroic doses of the same drug.

If theories of derivation could only stop here, content with explaining the diversification and succession of species between the tertiary period and the present time, through natural agencies or secondary causes still in operation, we fancy they would not be generally or violently objected to by the *savants* of the present day. But it is hard, if not impossible, to find a stopping-place. Some of the facts or accepted conclusions already referred to, and several others, of a more general character, which must be taken into the account, impel the theory onward with accumulated force. *Vires* (not to say *virus*) *acquirit eundo*. The theory hitches on wonderfully well to Lyell's uniformitarian theory in geology—that the thing that has been is the thing that is and shall be—that the natural operations now going on will account for all geological changes in a quiet and easy way, only give them time enough, so connecting the present and the proximate with the farthest past by almost imperceptible gradations—a view which finds large and increasing, if not general, acceptance in physical geology, and of which Darwin's theory is the natural complement.

So the Darwinian theory, once getting a foothold, marches boldly on, follows the supposed near ancestors of our present species farther and yet farther back into the dim past, and ends with an analogical inference which "makes the whole world kin." As we said at the beginning, this upshot discomposes us. Several features of the theory have an uncanny look. They may prove to be innocent: but their first aspect is suspicious, and high authorities pronounce the whole thing to be positively mischievous. In this dilemma we are going to take advice. Following the bent of our prejudices, and hoping to fortify these by new and strong arguments, we are going now to read the principal reviews which undertake to demolish the theory—with what result our readers shall be duly informed.

II.

"I can entertain no doubt, after the most deliberate study and dispassionate judgment of which I am capable, that the view which most naturalists entertain, and which I formerly entertained, namely, that each species has been independently created, is erroneous. I am fully convinced that species are not immutable; but that those belonging to what are called the same genera are lineal descendants of some other and generally extinct species, in the same manner as the acknowledged varieties of any one species are the descendants of that species. Furthermore, I am convinced that Natural Selection has been the main, but not exclusive, means of modification."

This is the kernel of the new theory, the Darwinian creed, as recited at the close of the introduction to the remarkable book under consideration. The questions, "What will he do with it?" and "How far will he carry it?" the author answers at the close of the volume:

"I cannot doubt that the theory of descent with modification embraces all the members of the same class." Furthermore, "I believe that all animals have descended from at most only four or five progenitors, and plants from an equal or lesser number."

Seeing that analogy as strongly suggests a further step in the same direction, while he protests that "analogy may be a deceitful guide," yet he follows its inexorable leading to the inference that—

"Probably all the organic beings which have ever lived on this earth have descended from some one primordial form, into which life was first breathed."[1]

[1] Page 484, English edition. In the new American edition (*vide* Supplement, pp. 431, 432) the principal analogies which suggest the

In the first extract we have the thin end of the wedge driven a little way; in the last, the wedge driven home.

We have already sketched some of the reasons suggestive of such a theory of derivation of species, reasons which gave it plausibility, and even no small probability, as applied to our actual world and to changes occurring since the latest tertiary period. We are well pleased at this moment to find that the conclusions we were arriving at in this respect are sustained by the very high authority and impartial judgment of Pictet, the Swiss paleontologist. In his review of Darwin's book[1]—the fairest and most admirable opposing one that has appeared—he freely accepts that *ensemble* of natural operations which Darwin impersonates under the now familiar name of Natural Selection, allows that the exposition throughout the first chapters seems "*à la fois prudent et fort*," and is disposed to accept the whole argument in its foundations, that is, so far as it relates to what is now going on, or has taken place in the present geological period—which period he carries back through the diluvial epoch to the borders of the tertiary.[2] Pictet accordingly admits that the

extreme view are referred to, and the remark is appended: "But this inference is chiefly grounded on analogy, and it is immaterial whether or not it be accepted. The case is different with the members of each great class, as the Vertebrata or Articulata; for here we have in the laws of homology, embryology, etc., some distinct evidence that all have descended from a single primordial parent."

[1] In *Bibliothèque Universelle de Genève*, March, 1860.

[2] This we learn from his very interesting article, "De la Question

theory will very well account for the origination by divergence of nearly-related species, whether within the present period or in remoter geological times; a very natural view for him to take, since he appears to have reached and published, several years ago, the pregnant conclusion that there most probably was some material connection between the closely-related species of two successive faunas, and that the numerous close species, whose limits are so difficult to determine, were not all created distinct and independent. But while thus accepting, or ready to accept, the basis of Darwin's theory, and all its legitimate direct inferences, he rejects the ultimate conclusions, brings some weighty arguments to bear against them, and is evidently convinced that he can draw a clear line between the sound inferences, which he favors, and the unsound or unwarranted theoretical deductions, which he rejects. We hope he can.

This raises the question, Why does Darwin press his theory to these extreme conclusions? Why do all hypotheses of derivation converge so inevitably to one ultimate point? Having already considered some of the reasons which suggest or support the theory at its outset—which may carry it as far as such sound and experienced naturalists as Pictet allow that it may be true—perhaps as far as Darwin himself unfolds it in the introductory proposition cited at the beginning of this article—we may now inquire after the

de l'Homme Fossile," in the same (March) number of the *Bibliothèque Universelle.* (*See,* also, the same author's "Note sur la Periode Quaternaire ou Diluvienne, considérée dans ses Rapports avec l'Époque Actuelle," in the number for August, 1860, of the same periodical.)

motives which impel the theorist so much farther. Here proofs, in the proper sense of the word, are not to be had. We are beyond the region of demonstration, and have only probabilities to consider. What are these probabilities? What work will this hypothesis do to establish a claim to be adopted in its completeness? Why should a theory which may plausibly enough account for the *diversification* of the species of each special type or genus be expanded into a general system for the *origination* or successive diversification of all species, and all special types or forms, from four or five remote primordial forms, or perhaps from one? We accept the theory of gravitation because it explains all the facts we know, and bears all the tests that we can put it to. We incline to accept the nebular hypothesis, for similar reasons; not because it is proved—thus far it is incapable of *proof*—but because it is a natural theoretical deduction from accepted physical laws, is thoroughly congruous with the facts, and because its assumption serves to connect and harmonize these into one probable and consistent whole. Can the derivative hypothesis be maintained and carried out into a system on similar grounds? If so, however unproved, it would appear to be a tenable hypothesis, which is all that its author ought now to claim. Such hypotheses as, from the conditions of the case, can neither be proved nor disproved by direct evidence or experiment, are to be tested only indirectly, and therefore imperfectly, by trying their power to harmonize the known facts, and to account for what is otherwise unaccountable. So the question comes to this: What will an hypothesis of the derivation of species explain which the opposing view leaves unexplained?

Questions these which ought to be entertained before we take up the arguments which have been advanced against this theory. We can barely glance at some of the considerations which Darwin adduces, or will be sure to adduce in the future and fuller exposition which is promised. To display them in such wise as to indoctrinate the unscientific reader would require a volume. Merely to refer to them in the most general terms would suffice for those familiar with scientific matters, but would scarcely enlighten those who are not. Wherefore let these trust the impartial Pictet, who freely admits that, "in the absence of sufficient direct proofs to justify the possibility of his hypothesis, Mr. Darwin relies upon indirect proofs, the bearing of which is real and incontestable;" who concedes that "his theory accords very well with the great facts of comparative anatomy and zoölogy—comes in admirably to explain unity of composition of organisms, also to explain rudimentary and representative organs, and the natural series of genera and species—equally corresponds with many paleontological data—agrees well with the specific resemblances which exist between two successive faunas, with the parallelism which is sometimes observed between the series of paleontological succession and of embryonal development," etc.; and finally, although he does not accept the theory in these results, he allows that "it appears to offer the best means of explaining the manner in which organized beings were produced in epochs anterior to our own."

What more than this could be said for such an hypothesis? Here, probably, is its charm, and its strong hold upon the speculative mind. Unproven though it be, and cumbered *prima facie* with cumulative improbabilities as it proceeds, yet it singularly accords with great classes of facts otherwise insulated and enigmatic, and explains many things which are thus far utterly inexplicable upon any other scientific assumption.

We have said that Darwin's hypothesis is the natural complement to Lyell's uniformitarian theory in physical geology. It is for the organic world what that is for the inorganic; and the accepters of the latter stand in a position from which to regard the former in the most favorable light. Wherefore the rumor that the cautious Lyell himself has adopted the Darwinian hypothesis need not surprise us. The two views are made for each other, and, like the two counterpart pictures for the stereoscope, when brought together, combine into one apparently solid whole.

If we allow, with Pictet, that Darwin's theory will very well serve for all that concerns the present epoch of the world's history—an epoch in which this renowned paleontologist includes the diluvial or quaternary period—then Darwin's first and foremost need in his onward course is a practicable road from this into and through the tertiary period, the intervening region between the comparatively near and the far remote past. Here Lyell's doctrine paves the way, by showing that in the physical geology there is no general or absolute break between the two, probably no greater between the latest tertiary and the quaternary period than between the latter and the present time. So far, the Lyellian view is, we suppose, generally concurred in. It is largely admitted that numerous tertiary species have continued down into the quaternary, and many of them to the present time. A goodly percentage of the earlier and nearly half of the later tertiary mollusca, according to Des Hayes, Lyell, and, if we mistake not, Bronn, still live. This identification, however, is now questioned by a naturalist of the very highest authority. But, in its bearings on the new theory, the point here turns not upon absolute identity so much as upon close resemblance. For those who, with Agassiz, doubt the specific identity in any of these cases, and those who say, with Pictet, that "the later tertiary deposits contain in general the *débris* of species *very nearly related* to those which still exist, belonging to the same genera, but specifically different," may also agree with Pictet, that the nearly-related species of successive faunas must or may have had "a material connection." But the only material connection that we have an idea of in such a case is a genealogical one. And the supposition of a genealogical connection is surely not unnatural in such cases—is demonstrably the natural one as respects all those tertiary species which experienced naturalists have pronounced to be identical with existing ones, but which others now deem distinct. For to identify the two is the same thing as to conclude the one to be the ancestor of the other. No doubt there are differences between the tertiary and the present individuals, differences equally noticed by both classes of naturalists, but differently estimated. By the one these are deemed quite compatible, by the other incompatible, with community of origin. *But who can tell us what amount of difference is compatible with community of origin?* This is the very question at issue, and one to be settled by observation alone. Who would have thought that the peach and the nectarine came from one stock? But, this being proved, is it now very improbable that both were derived from the almond, or from some common amygdaline progenitor? Who would have thought that the cabbage, cauliflower, broccoli, kale, and kohlrabi, are derivatives of one species, and rape or colza, turnip, and probably ruta-baga, of another species? And who that is convinced of this can long undoubtingly hold the original distinctness of turnips from cabbages as an article of faith? On scientific grounds may not a primordial cabbage or rape be assumed as the ancestor of all the cabbage races, on much the same ground that we assume a common ancestry for the diversified human races? If all our breeds of cattle came from one stock, why not this stock from the auroch, which has had all the time between the diluvial and the historic periods in which to set off a variation perhaps no greater than the difference between some sorts of domestic cattle?

That considerable differences are often discernible between tertiary individuals and their supposed descendants of the present day affords no argument against Darwin's theory, as has been rashly thought, but is decidedly in its favor. If the identification were so perfect that no more differences were observable between the tertiary and the recent shells than between various individuals of either, then Darwin's opponents, who argue the immutability of species from the ibises and cats preserved by the ancient Egyptians being just like those of the present day, could triumphantly add a few hundred thousand years more to the length of the experiment and to the force of their argument.

As the facts stand, it appears that, while some tertiary forms are essentially undistinguishable from existing ones, others are the same with a difference, which is judged not to be specific or aboriginal; and yet others show somewhat greater differences, such as are scientifically expressed by calling them marked varieties, or else doubtful species; while others, differing a little more, are confidently termed distinct, but nearly-related species. Now, is not all this a question of degree, of mere gradation of difference? And is it at all likely that these several gradations came to be established in two totally different ways—some of them (though naturalists can't agree which) through natural variation, or other secondary cause, and some by original creation, without secondary cause? We have seen that the judicious Pictet answers such questions as Darwin would have him do, in affirming that, in all probability, the nearly-related species of two successive faunas were materially connected, and that contemporaneous species, similarly resembling each other, were not all created so, but have become so. This is equivalent to saying that species (using the term as all naturalists do, and must continue to employ the word) have only a relative, not an absolute fixity; that differences fully equivalent to what are held to be specific may arise in the course of time, so that one species may at length be naturally replaced by another species a good deal like it, or may be diversified into two, three, or more species, or forms as different as species. This concedes all that Darwin has a right to ask, all that he can directly infer from evidence. We must add that it affords a *locus standi*, more or less tenable, for inferring more.

Here another geological consideration comes in to help on this inference. The species of the later tertiary period for the most part not only resembled those of our days—many of them so closely as to suggest an absolute continuity—but also occupied in general the same regions that their relatives occupy now. The same may be said, though less specially, of the earlier tertiary and of the later secondary; but there is less and less localization of forms as we recede, yet some localization even in palæozoic times. While in the secondary period one is struck with the similarity of forms and the identity of many of the species which flourished apparently at the same time in all or in the most widely-separated parts of the world, in the tertiary epoch, on the contrary, along with the increasing specialization of climates and their approximation to the present state, we find abundant evidence of increasing localization of orders, genera, and species; and this localization strikingly accords with the present geographical distribution of the same groups of species. Where the imputed forefathers lived, their relatives and supposed descendants now flourish. All the actual classes of the animal and vegetable kingdoms were represented in the tertiary faunas and floras, and in nearly the same proportions and the same diversities as at present. The faunas of what is now Europe, Asia, America, and Australia, differed from each other much as they now differ: in fact—according to Adolphe Brongniart, whose statements we here condense[1]—the inhabitants of these different regions appear for the most part to have acquired, before the close of the tertiary period, the characters which essentially distinguish their existing faunas. The Eastern Continent had then, as now, its great pachyderms, elephants, rhinoceros, hippopotamus; South America, its armadillos, sloths, and anteaters; Australia, a crowd of marsupials; and the very strange birds of New Zealand had predecessors of similar strangeness. Everywhere the same geographical distribution as now, with a difference in the particular area, as respects the northern portion of the continents, answering to a warmer climate then than ours, such as allowed species of hippopotamus, rhinoceros, and elephant, to range even to the regions now inhabited by the reindeer and the musk-ox, and with the serious disturbing intervention of the glacial period within a comparatively recent time. Let it be noted also that those tertiary species which have continued with little change down to our days are the marine animals of the lower grades, especially mollusca. Their low organization, moderate sensibility, and the simple conditions of an existence in a medium like the ocean, not subject to great variation and incapable of sudden change, may well account for their continuance; while, on the other hand, the more intense, however

[1] In *Comptes Rendus, Académie des Sciences*, February 2, 1857.

gradual, climatic vicissitudes on land, which have driven all tropical and subtropical forms out of the higher latitudes and assigned to them their actual limits, would be almost sure to extinguish such huge and unwieldy animals as mastodons, mammoths, and the like, whose power of enduring altered circumstances must have been small.

This general replacement of the tertiary species of a country by others so much like them is a noteworthy fact. The hypothesis of the independent creation of all species, irrespective of their antecedents, leaves this fact just as mysterious as is creation itself; that of derivation undertakes to account for it. Whether it satisfactorily does so or not, it must be allowed that the facts well accord with that hypothesis. The same may be said of another conclusion, namely, that the geological succession of animals and plants appears to correspond in a general way with their relative standing or rank in a natural system of classification. It seems clear that, though no one of the *grand types* of the animal kingdom can be traced back farther than the rest, yet the lower *classes* long preceded the higher; that there has been on the whole a steady progression within each class and order; and that the highest plants and animals have appeared only in relatively modern times. It is only, however, in a broad sense that this generalization is now thought to hold good. It encounters many apparent exceptions, and sundry real ones. So far as the rule holds, all is as it should be upon an hypothesis of derivation.

The rule has its exceptions. But, curiously enough,

the most striking class of exceptions, if such they be, seems to us even more favorable to the doctrine of derivation than is the general rule of a pure and simple ascending gradation. We refer to what Agassiz calls prophetic and synthetic types; for which the former name may suffice, as the difference between the two is evanescent.

"It has been noticed," writes our great zoölogist, "that certain types, which are frequently prominent among the representatives of past ages, combine in their structure peculiarities which at later periods are only observed separately in different, distinct types. Sauroid fishes before reptiles, Pterodactyles before birds, Ichthyosauri before dolphins, etc. There are entire families, of nearly every class of animals, which in the state of their perfect development exemplify such prophetic relations. . . . The sauroid fishes of the past geological ages are an example of this kind. These fishes, which preceded the appearance of reptiles, present a combination of ichthyic and reptilian characters not to be found in the true members of this class, which form its bulk at present. The Pterodactyles, which preceded the class of birds, and the Ichthyosauri, which preceded the Cetacea, are other examples of such prophetic types."—(Agassiz, "Contributions, Essay on Classification," p. 117.)

Now, these reptile-like fishes, of which gar-pikes are the living representatives, though of earlier appearance, are admittedly of higher rank than common fishes. They dominated until reptiles appeared, when they mostly gave place to (or, as the derivationists will insist, were resolved by divergent variation and natural selection into) common fishes, destitute of reptilian characters, and saurian reptiles—the intermediate grades, which, according to a familiar piscine saying, are "neither fish, flesh, nor good red-herring," being eliminated and extinguished by natural consequence of the struggle for existence which Darwin so aptly portrays. And so, perhaps, of the other prophetic types. Here type and antitype correspond. If these are true prophecies, we need not wonder that some who read them in Agassiz's book will read their fulfillment in Darwin's.

Note also, in this connection, that along with a wonderful persistence of type, with change of species, genera, orders, etc., from formation to formation, no species and no higher group which has once unequivocally died out ever afterward reappears. Why is this, but that the link of generation has been sundered? Why, on the hypothesis of independent originations, were not failing species recreated, either identically or with a difference, in regions eminently adapted to their well-being? To take a striking case. That no part of the world now offers more suitable conditions for wild horses and cattle than the pampas and other plains of South America, is shown by the facility with which they have there run wild and enormously multiplied, since introduced from the Old World not long ago. There was no wild American stock. Yet in the times of the mastodon and megatherium, at the dawn of the present period, wild-horses—certainly very much like the existing horse—roamed over those plains in abundance. On the principle of original and direct created adaptation of species to climate and other conditions, why were they not reproduced, when, after the colder intervening era, those regions became again eminently adapted to such animals? Why, but because, by their complete extinction in South America, the line of descent was there utterly broken? Upon the ordinary hypothesis, there is no scientific explanation possible of this series of facts, and of many others like them. Upon the new hypothesis, "the succession of the same types of structure within the same areas during the later geological periods ceases to be mysterious, and is simply explained by inheritance." Their cessation is failure of issue.

Along with these considerations the fact (alluded to on page 98) should be remembered that, as a general thing, related species of the present age are geographically associated. The larger part of the plants, and still more of the animals, of each separate country are peculiar to it; and, as most species now flourish over the graves of their by-gone relatives of former ages, so they now dwell among or accessibly near their kindred species.

Here also comes in that general "parallelism between the order of succession of animals and plants in geological times, and the gradation among their living representatives" from low to highly organized, from simple and general to complex and specialized forms; also "the parallelism between the order of succession of animals in geological times and the changes their living representatives undergo during their embryological growth," as if the world were one prolonged gestation. Modern science has much insisted on this parallelism, and to a certain extent is allowed to have made it out. All these things, which conspire to prove that the ancient and the recent forms of life "are somehow intimately connected together in one grand system," equally conspire to suggest that the connection is one similar or analogous to generation. Surely no naturalist can be blamed for entering somewhat confidently upon a field of speculative inquiry which here opens so invitingly; nor need former premature endeavors and failures utterly dishearten him.

All these things, it may naturally be said, go to explain the order, not the mode, of the incoming of species. But they all do tend to bring out the generalization expressed by Mr. Wallace in the formula that "every species has come into existence coïncident both in time and space with preëxisting closely-allied species." Not, however, that this is proved even of existing species as a matter of general fact. It is obviously impossible to *prove* anything of the kind. But we must concede that the known facts strongly suggest such an inference. And—since species are only congeries of individuals, since every individual came into existence in consequence of preëxisting individuals of the same sort, so leading up to the individuals with which the species began, and since the only material sequence we know of among plants and animals is that from parent to progeny—the presumption becomes exceedingly strong that the connection of the incoming with the preëxisting species is a genealogical one.

Here, however, all depends upon the probability that Mr. Wallace's inference is really true. Certainly it is not yet generally accepted; but a strong current is setting toward its acceptance.

So long as universal cataclysms were in vogue, and all life upon the earth was thought to have been suddenly destroyed and renewed many times in succession, such a view could not be thought of. So the equivalent view maintained by Agassiz, and formerly, we believe, by D'Orbigny, that irrespectively of general and sudden catastrophes, or any known adequate physical cause, there has been a total depopulation at the close of each geological period or formation, say forty or fifty times or more, followed by as many independent great acts of creation, at which alone have species been originated, and at each of which a vegetable and an animal kingdom were produced entire and complete, full-fledged, as flourishing, as wide-spread, and populous, as varied and mutually adapted from the beginning as ever afterward—such a view, of course, supersedes all material connection between successive species, and removes even the association and geographical range of species entirely out of the domain of physical causes and of natural science. This is the extreme opposite of Wallace's and Darwin's view, and is quite as hypothetical. The nearly universal opinion, if we rightly gather it, manifestly is, that the replacement of the species of successive formations was not complete and simultaneous, but partial and successive; and that along the course of each epoch some species probably were introduced, and some, doubtless, became extinct. If all since the tertiary belongs to our present epoch, this is certainly true of it: if to two or more epochs, then the hypothesis of a total change is not true of them.

Geology makes huge demands upon time; and we regret to find that it has exhausted ours—that what we meant for the briefest and most general sketch of some geological considerations in favor of Darwin's hypothesis has so extended as to leave no room for considering "the great facts of comparative anatomy and zoölogy" with which Darwin's theory "very well accords," nor for indicating how "it admirably serves for explaining the unity of composition of all organisms, the existence of representative and rudimentary organs, and the natural series which genera and species compose." Suffice it to say that these are the real strongholds of the new system on its theoretical side; that it goes far toward explaining both the physiological and the structural gradations and relations between the two kingdoms, and the arrangement of all their forms in groups subordinate to groups, all within a few great types; that it reads the riddle of abortive organs and of morphological conformity, of which no other theory has ever offered a scientific explanation, and supplies a ground for harmonizing the two fundamental ideas which naturalists and philosophers conceive to have ruled the organic world, though they could not reconcile them; namely, Adaptation to Purpose and Conditions of Existence, and Unity of Type. To reconcile these two undeniable principles is the capital problem in the philosophy of natural history; and the hypothesis which consistently does so thereby secures a great advantage.

We all know that the arm and hand of a monkey, the foreleg and foot of a dog and of a horse, the wing of a bat, and the fin of a porpoise, are fundamentally identical; that the long neck of the giraffe has the same and no more bones than the short one of the elephant; that the eggs of Surinam frogs hatch into tadpoles with as good tails for swimming as any of their kindred, although as tadpoles they never enter the water; that the Guinea-pig is furnished with incisor teeth which it never uses, as it sheds them before birth; that embryos of mammals and birds have branchial slits and arteries running in loops, in imitation or reminiscence of the arrangement which is permanent in fishes; and that thousands of animals and plants have rudimentary organs which, at least in numerous cases, are wholly useless to their possessors, etc., etc. Upon a derivative theory this morphological conformity is explained by community of descent; and it has not been explained in any other way.

Naturalists are constantly speaking of "related species," of the "affinity" of a genus or other group, and of "family resemblance"—vaguely conscious that these terms of kinship are something more than mere metaphors, but unaware of the grounds of their aptness. Mr. Darwin assures them that they have been talking derivative doctrine all their lives—as M. Jourdain talked prose—without knowing it.

If it is difficult and in many cases practically impossible to fix the limits of species, it is still more so to fix those of genera; and those of tribes and families are still less susceptible of exact natural circumscription. Intermediate forms occur, connecting one group with another in a manner sadly perplexing to systematists, except to those who have ceased to expect absolute limitations in Nature. All this blending could hardly fail to suggest a former material connection among allied forms, such as that which the hypothesis of derivation demands.

Here it would not be amiss to consider the general principle of gradation throughout organic Nature—a principle which answers in a general way to the Law of Continuity in the inorganic world, or rather is so analogous to it that both may fairly be expressed by the Leibnitzian axiom, *Natura non agit saltatim.* As an axiom or philosophical principle, used to test modal laws or hypotheses, this in strictness belongs only to physics. In the investigation of Nature at large, at least in the organic world, nobody would undertake to apply this principle as a test of the validity of any theory or supposed law. But naturalists of enlarged views will not fail to infer the principle from the phenomena they investigate—to perceive that the rule holds, under due qualifications and altered forms, throughout the realm of Nature; although we do not suppose that Nature in the organic world makes no distinct steps, but only short and serial steps—not infinitely fine gradations, but no long leaps, or few of them.

To glance at a few illustrations out of many that present themselves. It would be thought that the distinction between the two organic kingdoms was broad and absolute. Plants and animals belong to two very different categories, fulfill opposite offices, and, as to the mass of them, are so unlike that the difficulty of the ordinary observer would be to find points of comparison. Without entering into details, which would fill an article, we may safely say that the difficulty with the naturalist is all the other way—that all these broad differences vanish one by one as we approach the lower confines of the two kingdoms, and that no *abso-*

lute distinction whatever is now known between them. It is quite possible that the same organism may be both vegetable and animal, or may be first the one and then the other. If some organisms may be said to be at first vegetables and then animals, others, like the spores and other reproductive bodies of many of the lower Algæ, may equally claim to have first a characteristically animal, and then an unequivocally vegetable existence. Nor is the gradation restricted to these simple organisms. It appears in general functions, as in that of reproduction, which is reducible to the same formula in both kingdoms, while it exhibits close approximations in the lower forms; also in a common or similar ground of sensibility in the lowest forms of both, a common faculty of effecting movements tending to a determinate end, traces of which pervade the vegetable kingdom—while, on the other hand, this indefinable principle, this vegetable

"Animula vagula, blandula,
Hospes comesque corporis,"

graduates into the higher sensitiveness of the lower class of animals. Nor need we hesitate to recognize the fine gradations from simple sensitiveness and volition to the higher instinctive and to the other psychical manifestations of the higher brute animals. The gradation is undoubted, however we may explain it.

Again, propagation is of one mode in the higher animals, of two in all plants; but vegetative propagation, by budding or offshoots, extends through the lower grades of animals. In both kingdoms there may be separation of the offshoots, or indifference in this respect, or continued and organic union with the parent stock; and this either with essential independence of the offshoots, or with a subordination of these to a common whole; or finally with such subordination and amalgamation, along with specialization of function, that the same parts, which in other cases can be regarded only as progeny, in these become only members of an individual.

This leads to the question of individuality, a subject quite too large and too recondite for present discussion. The conclusion of the whole matter, however, is, that individuality—that very ground of *being* as distinguished from *thing*—is not attained in Nature at one leap. If anywhere truly exemplified in plants, it is only in the lowest and simplest, where the being is a structural unit, a single cell, memberless and organless, though organic—the same thing as those cells of which all the more complex plants are built up, and with which every plant and (structurally) every animal began its development. In the ascending gradation of the vegetable kingdom individuality is, so to say, striven after, but never attained; in the lower animals it is striven after with greater though incomplete success; it is realized only in animals of so high a rank that vegetative multiplication or offshoots are out of the question, where all parts are strictly members and nothing else, and all subordinated to a common nervous centre—is fully realized only in a conscious person.

So, also, the broad distinction between reproduction by seeds or ova and propagation by buds, though perfect in some of the lowest forms of life, becomes evanescent in others; and even the most absolute law we know in the physiology of genuine reproduction—that of sexual coöperation—has its exceptions in both kingdoms in parthenogenesis, to which in the vegetable kingdom a most curious and intimate series of gradations leads. In plants, likewise, a long and finely-graduated series of transitions leads from bisexual to unisexual blossoms; and so in various other respects. Everywhere we may perceive that Nature secures her ends, and makes her distinctions on the whole manifest and real, but everywhere without abrupt breaks. We need not wonder, therefore, that gradations between species and varieties should occur; the more so, since genera, tribes, and other groups into which the naturalist collocates species, are far from being always absolutely limited in Nature, though they are necessarily represented to be so in systems. From the necessity of the case, the classifications of the naturalist abruptly define where Nature more or less blends. Our systems are nothing, if not definite. They express differences, and some of the coarser gradations. But this evinces not their perfection, but their imperfection. Even the best of them are to the system of Nature what consecutive patches of the seven colors are to the rainbow.

Now the principle of gradation throughout organic Nature may, of course, be interpreted upon other assumptions than those of Darwin's hypothesis—certainly upon quite other than those of a materialistic philosophy, with which we ourselves have no sympathy. Still we conceive it not only possible, but probable, that this gradation, as it has its natural ground, may yet have its scientific explanation. In any case, there is no need to deny that the general facts correspond well with an hypothesis like Darwin's, which is built upon fine gradations.

We have contemplated quite long enough the general presumptions in favor of an hypothesis of the derivation of species. We cannot forget, however, while for the moment we overlook, the formidable difficulties which all hypotheses of this class have to encounter, and the serious implications which they seem to involve. We feel, moreover, that Darwin's particular hypothesis is exposed to some special objections. It requires no small strength of nerve steadily to conceive, not only of the diversification, but of the formation of the organs of an animal through cumulative variation and natural selection. Think of such an organ as the eye, that most perfect of optical instruments, as so produced in the lower animals and perfected in the higher! A friend of ours, who accepts the new doctrine, confesses that for a long while a cold chill came over him whenever he thought of the eye. He has at length got over that stage of the complaint, and is now in the fever of belief, perchance to be succeeded by the sweating stage, during which sundry peccant humors may be eliminated from the system. For ourselves, we dread the chill, and have some misgiving about the consequences of the reaction. We find ourselves in the "singular position" acknowledged by Pictet—that is, confronted with a theory which, although it can really explain much, seems inadequate to the heavy task it so boldly assumes, but which, nevertheless, appears better fitted than any other that has been broached to explain, if it be possible to explain, somewhat of the manner in which organized beings may have arisen and succeeded each other. In this dilemma we might take advantage of Mr. Darwin's candid admission, that he by no means expects to convince old and experienced people, whose minds are stocked with a multitude of facts all regarded during a long course of years from the old point of view. This is nearly our case. So, owning no call to a larger faith than is expected of us, but not prepared to pronounce the whole hypothesis untenable, under such construction as we should put upon it, we naturally sought to attain a settled conviction through a perusal of several proffered refutations of the theory. At least, this course seemed to offer the readiest way of bringing to a head the various objections to which the theory is exposed. On several accounts some of these opposed reviews especially invite examination. We propose, accordingly, to conclude our task with an article upon "Darwin and his Reviewers."

III.

THE origin of species, like all origination, like the institution of any other natural state or order, is beyond our immediate ken. We see or may learn how things go on; we can only frame hypotheses as to how they began.

Two hypotheses divide the scientific world, very unequally, upon the origin of the existing diversity of the plants and animals which surround us. One assumes that the actual kinds are primordial; the other, that they are derivative. One, that all kinds originated supernaturally and directly as such, and have continued unchanged in the order of Nature; the other, that the present kinds appeared in some sort of genealogical connection with other and earlier kinds, that they became what they now are in the course of time and in the order of Nature.

Or, bringing in the word *species*, which is well defined as "the perennial succession of individuals," commonly of very like individuals—as a close corporation of individuals perpetuated by generation, instead of election—and reducing the question to mathematical simplicity of statement: species are lines of individuals coming down from the past and running on to the future; lines receding, therefore, from our view in either direction. Within our limited observation they appear to be parallel lines, as a general thing neither approaching to nor diverging from each other.

The first hypothesis assumes that they were parallel from the unknown beginning and will be to the unknown end. The second hypothesis assumes that the apparent parallelism is not real and complete, at least aboriginally, but approximate or temporary; that we should find the lines convergent in the past, if we could trace them far enough; that some of them, if produced back, would fall into certain fragments of lines, which have left traces in the past, lying not exactly in the same direction, and these farther back into others to which they are equally unparallel. It will also claim that the present lines, whether on the whole really or only approximately parallel, sometimes fork or send off branches on one side or the other, producing new lines (varieties), which run for a while, and for aught we know indefinitely when not interfered with, near and approximately parallel to the parent line. This claim it can establish; and it may also show that these close subsidiary lines may branch or vary again, and that those branches or varieties which are best adapted to the existing conditions may be continued, while others stop or die out. And so we may have the basis of a real *theory* of the *diversification* of species; and here, indeed, there is a real, though a narrow, established ground to build upon. But, as systems of organic Nature, both doctrines are equally *hypotheses*, are suppositions of what there is no proof of from experience, assumed in order to account for the observed phenomena, and supported by such indirect evidence as can be had.

Even when the upholders of the former and more popular system mix up revelation with scientific discussion—which we decline to do—they by no means thereby render their view other than hypothetical. Agreeing that plants and animals were produced by Omnipotent fiat does not exclude the idea of natural order and what we call secondary causes. The record of the fiat—"Let the earth bring forth grass, the herb yielding seed," etc., "and it was so;" "let the earth bring forth the living creature after his kind, cattle and creeping thing and beast of the earth after his kind, and it was so"—seems even to imply them. Agreeing that they were formed of "the dust of the ground," and of thin air, only leads to the conclusion that the pristine individuals were corporeally constituted like existing individuals, produced through natural agencies. To agree that they were created "after their kinds" determines nothing as to what were the original kinds, nor in what mode, during what time, and in what connections it pleased the Almighty to introduce the first individuals of each sort upon the earth. Scientifically considered, the two opposing doctrines are equally hypothetical.

The two views very unequally divide the scientific world; so that believers in "the divine right of majorities" need not hesitate which side to take, at least for the present. Up to a time quite within the memory of a generation still on the stage, two hypotheses about the nature of light very unequally divided the scientific world. But the small minority has already prevailed: the emission theory has gone out; the undulatory or wave theory, after some fluctuation, has reached high tide, and is now the pervading, the fully-established system. There was an intervening time during which most physicists held their opinions in suspense.

The adoption of the undulatory theory of light called for the extension of the same theory to heat, and this promptly suggested the hypothesis of a correlation, material connection, and transmutability of heat, light, electricity, magnetism, etc.; which hypothesis the physicists held in absolute suspense until very lately, but are now generally adopting. If not already established as a system, it promises soon to become so. At least, it is generally received as a tenable and probably true hypothesis.

Parallel to this, however less cogent the reasons, Darwin and others, having shown it likely that some varieties of plants or animals have diverged in time into cognate species, or into forms as different as species, are led to infer that all species of a genus may have thus diverged from a common stock, and thence to suppose a higher community of origin in ages still farther back, and so on. Following the safe example of the physicists, and acknowledging the fact of the diversification of a once homogeneous species into varieties, we may receive the theory of the evolution of these into species, even while for the present we hold the hypothesis of a further evolution in cool suspense or in grave suspicion. In respect to very many questions a wise man's mind rests long in a state neither of belief nor of unbelief. But your intellectually short-sighted people are apt to be preternaturally clear-sighted, and to find their way very plain to positive conclusions upon one side or the other of every mooted question.

In fact, most people, and some philosophers, refuse to hold questions in abeyance, however incompetent they may be to decide them. And, curiously enough, the more difficult, recondite, and perplexing, the questions or hypotheses are—such, for instance, as those about organic Nature—the more impatient they are of suspense. Sometimes, and evidently in the present case, this impatience grows out of a fear that a new hypothesis may endanger cherished and most important beliefs. Impatience under such circumstances is not unnatural, though perhaps needless, and, if so, unwise.

To us the present revival of the derivative hypothesis, in a more winning shape than it ever before had, was not unexpected. We wonder that any thoughtful observer of the course of investigation and of speculation in science should not have foreseen it, and have learned at length to take its inevitable coming patiently; the more so, as in Darwin's treatise it comes in a purely scientific form, addressed only to scientific men. The notoriety and wide popular perusal of this treatise appear to have astonished the author even more than the book itself has astonished the reading world. Coming, as the new presentation does, from a naturalist of acknowledged character and ability, and marked by a conscientiousness and candor which have not always been reciprocated, we have thought it simply right to set forth the doctrine as fairly and as favorably as we could. There are plenty to decry it, and the whole theory is widely exposed to attack. For the arguments on the other side we may look to the numerous adverse publications which Darwin's volume has already called out, and especially to those reviews which propose directly to refute it. Taking various lines, and reflecting very diverse modes of thought, these hostile critics may be expected to concentrate and enforce the principal objections which can be brought to bear against the derivative hypothesis in general, and Darwin's new exposition of it in particular.

Upon the opposing side of the question we have read with attention—1. An article in the *North American Review* for April last; 2. One in the *Christian Examiner*, Boston, for May; 3. M. Pictet's article in the *Bibliothèque Universelle*, which we have already made considerable use of, which seems throughout most able and correct, and which in tone and fairness is admirably in contrast with—4. The article in the *Edinburgh Review* for May, attributed—although against a large amount of internal presumptive evidence—to the most distinguised British comparative anatomist; 5. An article in the *North British Review* for May; 6. Prof. Agassiz has afforded an early opportunity to peruse the criticisms he makes in the forthcoming third volume of his great work, by a publication of them in advance in the *American Journal of Science* for July.

In our survey of the lively discussion which has been raised, it matters little how our own particular opinions may incline. But we may confess to an impression, thus far, that the doctrine of the permanent and complete immutability of species has not been established, and may fairly be doubted. We believe that species vary, and that "Natural Selection" works; but we suspect that its operation, like every analogous natural operation, may be limited by something else. Just as every species by its natural rate of reproduction would soon completely fill any country it could live in, but does not, being checked by some other species or some other condition—so it may be surmised that variation and natural selection have their struggle and consequent check, or are limited by something inherent in the constitution of organic beings.

We are disposed to rank the derivative hypothesis in its fullness with the nebular hypothesis, and to regard both as allowable, as not unlikely to prove tenable in spite of some strong objections, but as not therefore demonstrably true. Those, if any there be, who regard the derivative hypothesis as satisfactorily proved, must have loose notions as to what proof is. Those who imagine it can be easily refuted and cast aside, must, we think, have imperfect or very prejudiced conceptions of the facts concerned and of the questions at issue.

We are not disposed nor prepared to take sides for or against the new hypothesis, and so, perhaps, occupy a good position from which to watch the discussion and criticise those objections which are seemingly inconclusive. On surveying the arguments urged by those who have undertaken to demolish the theory, we have been most impressed with a sense of their great inequality. Some strike us as excellent and perhaps unanswerable; some, as incongruous with other views of the same writers; others, when carried out, as incompatible with general experience or general beliefs, and therefore as proving too much; still others, as proving nothing at all; so that, on the whole, the effect is rather confusing and disappointing. We certainly expected a stronger adverse case than any which the thoroughgoing opposers of Darwin appear to have made out. Wherefore, if it be found that the new hypothesis has grown upon our favor as we proceeded, this must be attributed not so much to the force of the arguments of the book itself as to the want of force of several of those by which it has been assailed. Darwin's arguments we might resist or adjourn; but some of the refutations of it give us more concern than the book itself did.

These remarks apply mainly to the philosophical and theological objections which have been elaborately urged, almost exclusively by the American reviewers. The *North British* reviewer, indeed, roundly denounces the book as atheistical, but evidently deems the case too clear for argument. The *Edinburgh* reviewer, on the contrary, scouts all such objections —as well he may, since he records his belief in "a continuous creative operation," a constantly operating secondary creational law," through which species are successively produced; and he emits faint, but not indistinct, glimmerings of a transmutation theory of his own;[1] so that he is equally exposed to all the

[1] Whatever it may be, it is not "the homœopathic form of the transmutative hypothesis," as Darwin's is said to be (p. 252, American reprint), so happily that the prescription is repeated in the second (p. 259) and third (p. 271) dilutions, no doubt, on Hahnemann's famous princi-

philosophical objections advanced by Agassiz, and to most of those urged by the other American critics, against Darwin himself.

Proposing now to criticise the critics, so far as to see what their most general and comprehensive objections amount to, we must needs begin with the American reviewers, and with their arguments adduced to prove that a derivative hypothesis *ought not to be true*, or is not possible, philosophical, or theistic.

It must not be forgotten that on former occasions very confident judgments have been pronounced by very competent persons, which have not been finally ratified. Of the two great minds of the seventeenth century, Newton and Leibnitz, both profoundly religious as well as philosophical, one produced the theory of gravitation, the other objected to that theory that it was subversive of natural religion. The nebular hypothesis—a natural consequence of the theory of gravitation and of the subsequent progress of physical and astronomical discovery—has been denounced as atheistical even down to our own day. But it is now largely adopted by the most theistical natural philosophers as a tenable and perhaps sufficient hypothesis, and where not accepted is no longer objected to, so far as we know, on philosophical or religious grounds.

The gist of the philosophical objections urged by

ple, of an increase of potency at each dilution. Probably the supposed transmutation is *per saltus*. "Homœopathic doses of transmutation," indeed! Well, if we really must swallow transmutation in some form or other, as this reviewer intimates, we might prefer the mild homœopathic doses of Darwin's formula to the allopathic bolus which the Edinburgh general practitioner appears to be compounding.

the two Boston reviewers against an hypothesis of the derivation of species—or at least against Darwin's particular hypothesis—is, that it is incompatible with the idea of any manifestation of design in the universe, that it denies final causes. A serious objection this, and one that demands very serious attention.

The proposition, that things and events in Nature were not designed to be so, if logically carried out, is doubtless tantamount to atheism. Yet most people believe that some were designed and others were not, although they fall into a hopeless maze whenever they undertake to define their position. So we should not like to stigmatize as atheistically disposed a person who regards certain things and events as being what they are through designed laws (whatever that expression means), but as not themselves specially ordained, or who, in another connection, believes in general, but not in particular Providence. We could sadly puzzle him with questions; but in return he might equally puzzle us. Then, to deny that anything was specially designed to be what it is, is one proposition; while to deny that the Designer supernaturally or immediately made it so, is another: though the reviewers appear not to recognize the distinction.

Also, "scornfully to repudiate" or to "sneer at the idea of any manifestation of design in the material universe,"[1] is one thing; while to consider, and perhaps to exaggerate, the difficulties which attend the practical application of the doctrine of final causes to

[1] Vide *North American Review*, for April, 1860, p. 475, and *Christian Examiner*, for May, p. 457.

certain instances, is quite another thing: yet the Boston reviewers, we regret to say, have not been duly regardful of the difference. Whatever be thought of Darwin's doctrine, we are surprised that he should be charged with *scorning* or *sneering* at the opinions of others, upon such a subject. Perhaps Darwin's view is incompatible with final causes—we will consider that question presently—but as to the *Examiner's* charge, that he "sneers at the idea of any manifestation of design in the material universe," though we are confident that no misrepresentation was intended, we are equally confident that it is not at all warranted by the two passages cited in support of it. Here are the passages:

> "If green woodpeckers alone had existed, or we did not know that there were many black and pied kinds, I dare say that we should have thought that the green color was a beautiful adaptation to hide this tree-frequenting bird from its enemies."
>
> "If our reason leads us to admire with enthusiasm a multitude of inimitable contrivances in Nature, this same reason tells us, though we may easily err on both sides, that some contrivances are less perfect. Can we consider the sting of the wasp or of the bee as perfect, which, when used against many attacking animals, cannot be withdrawn, owing to the backward serratures, and so inevitably causes the death of the insect by tearing out its viscera?"

If the sneer here escapes ordinary vision in the detached extracts (one of them wanting the end of the sentence), it is, if possible, more imperceptible when read with the context. Moreover, this perusal inclines us to think that the *Examiner* has misapprehended the particular argument or object, as well as the spirit, of the author in these passages. The whole reads more naturally as a caution against the inconsiderate use of final causes in science, and an illustration of some of the manifold errors and absurdities which their hasty assumption is apt to involve—considerations probably equivalent to those which induced Lord Bacon to liken final causes to "vestal virgins." So, if any one, it is here Bacon that "sitteth in the seat of the scornful." As to Darwin, in the section from which the extracts were made, he is considering a subsidiary question, and trying to obviate a particular difficulty, but, we suppose, is wholly unconscious of denying "any manifestation of design in the material universe." He concludes the first sentence:

—"and consequently that it was a character of importance, and might have been acquired through natural selection; as it is, I have no doubt that the color is due to some quite distinct cause, probably to sexual selection."

After an illustration from the vegetable creation, Darwin adds:

"The naked skin on the head of a vulture is generally looked at as a *direct* adaptation for wallowing in putridity; *and so it may be*, or it may possibly be due to the direct action of putrid matter; but we should be very cautious in drawing any such inference, when we see that the skin on the head of the clean-feeding male turkey is likewise naked. The sutures in the skulls of young mammals have been advanced as a beautiful adaptation for aiding parturition, and no doubt they facilitate or may be indispensable for this act; but as sutures occur in the skulls of young birds and reptiles, which have only to escape from a broken egg, we may infer that this structure has arisen from the laws of growth, and has been taken advantage of in the parturition of the higher animals."

All this, simply taken, is beyond cavil, unless the attempt to explain scientifically how any designed result is accomplished savors of impropriety.

In the other place, Darwin is contemplating the patent fact that "perfection here below" is relative, not absolute—and illustrating this by the circumstance that European animals, and especially plants, are now proving to be better adapted for New Zealand than many of the indigenous ones—that "the correction for the aberration of light is said, on high authority, not to be quite perfect even in that most perfect organ, the eye." And then follows the second extract of the reviewer. But what is the position of the reviewer upon his own interpretation of these passages? If he insists that green woodpeckers were specifically created so in order that they might be less liable to capture, must he not equally hold that the black and pied ones were specifically made of these colors in order that they might be more liable to be caught? And would an explanation of the mode in which those woodpeckers came to be green, however complete, convince him that the color was undesigned?

As to the other illustration, is the reviewer so complete an optimist as to insist that the arrangement and the weapon are wholly perfect (*quoad* the insect) the normal use of which often causes the animal fatally to injure or to disembowel itself? Either way it seems to us that the argument here, as well as the insect, performs *hari-kari*. The *Examiner* adds:

"We should in like manner object to the word *favorable*, as implying that some species are placed by the Creator under *unfavorable* circumstances, at least under such as might be advantageously modified."

But are not many individuals and some races of men placed by the Creator "under unfavorable circumstances, at least under such as might be advantageously modified?" Surely these reviewers must be living in an ideal world, surrounded by "the faultless monsters which *our* world ne'er saw," in some elysium where imperfection and distress were never heard of! Such arguments resemble some which we often hear against the Bible, holding that book responsible as if it originated certain facts on the shady side of human nature or the apparently darker lines of Providential dealing, though the facts are facts of common observation and have to be confronted upon any theory.

The *North American* reviewer also has a world of his own—just such a one as an idealizing philosopher would be apt to devise—that is, full of sharp and absolute distinctions: such, for instance, as the "absolute invariableness of instinct;" an absolute want of intelligence in any brute animal; and a complete monopoly of instinct by the brute animals, so that this "instinct is a great matter" for them only, since it sharply and perfectly distinguishes this portion of organic Nature from the vegetable kingdom on the one hand and from man on the other: most convenient views for argumentative purposes, but we suppose not borne out in fact.

In their scientific objections the two reviewers take somewhat different lines; but their philosophical and theological arguments strikingly coincide. They agree in emphatically asserting that Darwin's hypothesis of the origination of species through variation and natural selection "repudiates the whole doctrine of final causes," and "all indication of design or purpose in the organic world. . . . is neither more nor less than a formal denial of any agency beyond that of a blind chance in the developing or perfecting of the organs or instincts of created beings. . . . It is in vain that the apologists of this hypothesis might say that it merely attributes a different mode and time to the Divine agency—that all the qualities subsequently appearing in their descendants must have been implanted, and have remained latent in the original pair." Such a view, the *Examiner* declares, "is nowhere stated in this book, and would be, we are sure, disclaimed by the author."

We should like to be informed of the grounds of this sureness. The marked rejection of spontaneous generation—the statement of a belief that all animals have descended from four or five progenitors, and plants from an equal or lesser number, or, perhaps, if constrained to it by analogy, "from some one primordial form into which life was first breathed"—coupled with the expression, "To my mind it accords better with what we know of the laws impressed on matter by the Creator, that the production and extinction of the past and present inhabitants of the world should have been due to secondary causes," than "that each species has been independently created"—these and similar expressions lead us to suppose that the author probably does accept the kind of view which the *Examiner* is sure he would disclaim. At least, we charitably see nothing in his scientific theory to hinder his adoption of Lord Bacon's "Confession of Faith" in this regard—

"That, notwithstanding God hath rested and ceased from creating [in the sense of supernatural origination], yet, nevertheless, he doth accomplish and fulfill his divine will in all things, great and small, singular and general, as fully and exactly by providence as he could by miracle and new creation, though his working be not immediate and direct, but by compass; not violating Nature, which is his own law upon the creature."

However that may be, it is undeniable that Mr. Darwin has purposely been silent upon the philosophical and theological applications of his theory. This reticence, under the circumstances, argues design, and raises inquiry as to the final cause or reason why. Here, as in higher instances, confident as we are that there is a final cause, we must not be over-confident that we can infer the particular or true one. Perhaps the author is more familiar with natural-historical than with philosophical inquiries, and, not having decided which particular theory about efficient cause is best founded, he meanwhile argues the scientific questions concerned—all that relates to secondary causes—upon purely scientific grounds, as he must do in any case. Perhaps, confident, as he evidently is, that his view will finally be adopted, he may enjoy a sort of satisfaction in hearing it denounced as sheer atheism by the inconsiderate, and afterward, when it takes its place with the nebular hypothesis and the like, see this judgment reversed, as we suppose it would be in such event.

Whatever Mr. Darwin's philosophy may be, or whether he has any, is a matter of no consequence at all, compared with the important questions, whether a theory to account for the origination and diversification of animal and vegetable forms through the operation of secondary causes does or does not exclude design; and whether the establishment by adequate evidence of Darwin's particular theory of diversification through variation and natural selection would essentially alter the present scientific and philosophical grounds for theistic views of Nature. The unqualified affirmative judgment rendered by the two Boston reviewers, evidently able and practised reasoners, "must give us pause." We hesitate to advance our conclusions in opposition to theirs. But, after full and serious consideration, we are constrained to say that, in our opinion, the adoption of a derivative hypothesis, and of Darwin's particular hypothesis, if we understand it, would leave the doctrines of final causes, utility, and special design, just where they were before. We do not pretend that the subject is not environed with difficulties. Every view is so environed; and every shifting of the view is likely, if it removes some difficulties, to bring others into prominence. But we cannot perceive that Darwin's theory brings in any new kind of scientific difficulty, that is, any with which philosophical naturalists were not already familiar.

Since natural science deals only with secondary or natural causes, the scientific terms of a theory of derivation of species—no less than of a theory of dynamics—must needs be the same to the theist as to the atheist. The difference appears only when the inquiry is carried up to the question of primary cause—a question which belongs to philosophy. Wherefore, Darwin's reticence about efficient cause does not disturb us. He considers only the scientific questions. As already stated, we think that a theistic view of Nature is implied in his book, and we must charitably refrain from suggesting the contrary until the contrary is logically deduced from his premises. If, however, he anywhere maintains that the natural causes through which species are diversified operate without an ordaining and directing intelligence, and that the orderly arrangements and admirable adaptations we see all around us are fortuitous or blind, undesigned results—that the eye, though it came to see, was not designed for seeing, nor the hand for handling—then, we suppose, he is justly chargeable with denying, and very needlessly denying, all design in organic Nature; otherwise, we suppose not. Why, if Darwin's well-known passage about the eye[1]—equivocal though some of the language be—does not imply ordaining and directing intelligence, then he refutes his own theory as effectually as any of his opponents are likely to do. He asks:

"May we not believe that [under variation proceeding long enough, generation multiplying the better variations times enough, and natural selection securing the improvements] a living optical instrument might be thus formed as superior to one of glass as the works of the Creator are to those of man?"

This must mean one of two things: either that the living instrument was made and perfected under (which is the same thing as by) an intelligent First Cause, or that it was not. If it was, then theism is asserted; and as to the mode of operation, how do we know, and why must we believe, that, fitting precedent forms being in existence, a living instrument (so different

[1] Page 188, English edition.

from a lifeless manufacture) would be originated and perfected in any other way, or that this is not the fitting way? If it means that it was not, if he so misuses words that by the Creator he intends an unintelligent power, undirected force, or necessity, then he has put his case so as to invite disbelief in it. For then blind forces have produced not only manifest adaptations of means to specific ends—which is absurd enough—but better adjusted and more perfect instruments or machines than intellect (that is, human intellect) can contrive and human skill execute—which no sane person will believe.

On the other hand, if Darwin even admits—we will not say adopts—the theistic view, he may save himself much needless trouble in the endeavor to account for the absence of every sort of intermediate form. Those in the line between one species and another supposed to be derived from it he may be bound to provide; but as to "an infinite number of other varieties not intermediate, gross, rude, and purposeless, the unmeaning creations of an unconscious cause," born only to perish, which a relentless reviewer has imposed upon his theory—rightly enough upon the atheistic alternative—the theistic view rids him at once of this "scum of creation." For, as species do not now vary at all times and places and in all directions, nor produce crude, vague, imperfect, and useless forms, there is no reason for supposing that they ever did. Good-for-nothing monstrosities, failures of purpose rather than purposeless, indeed, sometimes occur; but these are just as anomalous and unlikely upon Darwin's theory as upon any other. For his particular theory is based, and even over-strictly insists, upon the most universal of physiological laws, namely, that successive generations shall differ only slightly, if at all, from their parents; and this effectively excludes crude and impotent forms. Wherefore, if we believe that the species were designed, and that natural propagation was designed, how can we say that the actual varieties of the species were not equally designed? Have we not similar grounds for inferring design in the supposed varieties of species, that we have in the case of the supposed species of a genus? When a naturalist comes to regard as three closely-related species what he before took to be so many varieties of one species, how has he thereby strengthened our conviction that the three forms are designed to have the differences which they actually exhibit? Wherefore, so long as gradatory, orderly, and adapted forms in Nature argue design, and at least while the physical cause of variation is utterly unknown and mysterious, we should advise Mr. Darwin to assume, in the philosophy of his hypothesis, that variation has been led along certain beneficial lines. Streams flowing over a sloping plain by gravitation (here the counterpart of natural selection) may have worn their actual channels as they flowed; yet their particular courses may have been assigned; and where we see them forming definite and useful lines of irrigation, after a manner unaccountable on the laws of gravitation and dynamics, we should believe that the distribution was designed.

To insist, therefore, that the new hypothesis of the derivative origin of the actual species is incompatible with final causes and design, is to take a position which we must consider philosophically untenable. We must also regard it as highly unwise and dangerous, in the present state and present prospects of physical and physiological science. We should expect the philosophical atheist or skeptic to take this ground; also, until better informed, the unlearned and unphilosophical believer; but we should think that the thoughtful theistic philosopher would take the other side. Not to do so seems to concede that only supernatural events can be shown to be designed, which no theist can admit—seems also to misconceive the scope and meaning of all ordinary arguments for design in Nature. This misconception is shared both by the reviewers and the reviewed. At least, Mr. Darwin uses expressions which imply that the natural forms which surround us, because they have a history or natural sequence, could have been only generally, but not particularly designed—a view at once superficial and contradictory; whereas his true line should be, that his hypothesis concerns the *order* and not the *cause*, the *how* and not the *why* of the phenomena, and so leaves the question of design just where it was before.

To illustrate this from the theist's point of view: Transfer the question for a moment from the origination of species to the origination of individuals, which occurs, as we say, naturally. Because natural, that is, "stated, fixed, or settled," is it any the less designed on that account? We acknowledge that God is our maker—not merely the originator of the race, but *our* maker as individuals—and none the less so because it pleased him to make us in the way of ordinary generation. If any of us were born unlike our parents and grandparents, in a slight degree, or in whatever degree, would the case be altered in this regard?

The whole argument in natural theology proceeds upon the ground that the inference for a final cause of the structure of the hand and of the valves in the veins is just as valid now, in individuals produced through natural generation, as it would have been in the case of the first man, supernaturally created. Why not, then, just as good even on the supposition of the descent of men from chimpanzees and gorillas, since those animals possess these same contrivances? Or, to take a more supposable case: If the argument from structure to design is convincing when drawn from a particular animal, say a Newfoundland dog, and is not weakened by the knowledge that this dog came from similar parents, would it be at all weakened if, in tracing his genealogy, it were ascertained that he was a remote descendant of the mastiff or some other breed, or that both these and other breeds came (as is suspected) from some wolf? If not, how is the argument for design in the structure of our particular dog affected by the supposition that his wolfish progenitor came from a post-tertiary wolf, perhaps less unlike an existing one than the dog in question is to some other of the numerous existing races of dogs, and that this post-tertiary came from an equally or more different tertiary wolf? And if the argument from structure to design is not invalidated by our present knowledge that our individual dog was developed from a single organic cell, how is it invalidated by the supposition of an analogous natural descent, through a long line of connected forms, from such a cell, or from some simple animal, existing ages before there were any dogs?

Again, suppose we have two well-known and apparently most decidedly different animals or plants, A and D, both presenting, in their structure and in their adaptations to the conditions of existence, as valid and clear evidence of design as any animal or plant ever presented: suppose we have now discovered two intermediate species, B and C, which make up a series with equable differences from A to D. Is the proof of design or final cause in A and D, whatever it amounted to, at all weakened by the discovery of the intermediate forms? Rather does not the proof extend to the intermediate species, and go to show that all four were equally designed? Suppose, now, the number of intermediate forms to be much increased, and therefore the gradations to be closer yet—as close as those between the various sorts of dogs, or races of men, or of horned cattle: would the evidence of design, as shown in the structure of any of the members of the series, be any weaker than it was in the case of A and D? Whoever contends that it would be, should likewise maintain that the origination of individuals by generation is incompatible with design, or an impossibility in Nature. We might all have confidently thought the latter, antecedently to experience of the fact of reproduction. Let our experience teach us wisdom.

These illustrations make it clear that the evidence of design from structure and adaptation is furnished *complete* by the individual animal or plant itself, and that our knowledge or our ignorance of the history of its formation or mode of production adds nothing to it and takes nothing away. We infer design from certain arrangements and results ; and we have no other way of ascertaining it. Testimony, unless infallible, cannot prove it, and is out of the question here. *Testimony is not the appropriate proof of design: adaptation to purpose is.* Some arrangements in Nature appear to be contrivances, but may leave us in doubt. Many others, of which the eye and the hand are notable examples, compel belief with a force not appreciably short of demonstration. Clearly to settle that such as these must have been designed goes far toward proving that other organs and other seemingly less explicit adaptations in Nature must also have been designed, and clinches our belief, from manifold considerations, that all Nature is a preconcerted arrangement, a manifested design. A strange contradiction would it be to insist that the shape and markings of certain rude pieces of flint, lately found in drift-deposits, prove design, but that nicer and thousand-fold more complex adaptations to use in animals and vegetables do not *a fortiori* argue design.

We could not affirm that the arguments for design in Nature are conclusive to all minds. But we may insist, upon grounds already intimated, that, whatever they were good for before Darwin's book appeared, they are good for now. To our minds the argument from design always appeared conclusive of the being and continued operation of an intelligent First Cause, the Ordainer of Nature; and we do not see that the grounds of such belief would be disturbed or shifted by the adoption of Darwin's hypothesis. We are not blind to the philosophical difficulties which the thoroughgoing implication of design in Nature has to encounter, nor is it our vocation to obviate them. It suffices us to know that they are not new nor peculiar difficulties—that, as Darwin's theory and our reasonings upon it did not raise these perturbing spirits, they are not bound to lay them. Meanwhile, that the doctrine of design encounters the very same difficulties in the material that it does in the moral world is just what ought to be expected.

So the issue between the skeptic and the theist is only the old one, long ago argued out—namely, whether organic Nature is a result of design or of chance. Variation and natural selection open no third alternative; they concern only the question how the results, whether fortuitous or designed, may have been brought about. Organic Nature abounds with unmistakable and irresistible indications of design, and, being a connected and consistent system, this evidence carries the implication of design throughout the whole. On the other hand, chance carries no probabilities with it, can never be developed into a consistent system, but, when applied to the explanation of orderly or beneficial results, heaps up improbabilities at every step beyond all computation. To us, a fortuitous Cosmos is simply inconceivable. The alternative is a designed Cosmos.

It is very easy to assume that, because events in Nature are in one sense accidental, and the operative forces which bring them to pass are themselves blind and unintelligent (physically considered, all forces are), therefore they are undirected, or that he who describes these events as the results of such forces thereby assumes that they are undirected. This is the assumption of the Boston reviewers, and of Mr. Agassiz, who insists that the only alternative to the doctrine, that all organized beings were supernaturally created just as they are, is, that they have arisen *spontaneously* through the *omnipotence of matter*.[1]

As to all this, nothing is easier than to bring out in the conclusion what you introduce in the premises. If you import atheism into your conception of variation and natural selection, you can readily exhibit it in the result. If you do not put it in, perhaps there need be none to come out. While the mechanician is considering a steamboat or locomotive-engine as a material organism, and contemplating the fuel, water, and steam, the source of the mechanical forces, and how they operate, he may not have occasion to mention the engineer. But, the orderly and special results accomplished, the *why* the movements are in this or that particular direction, etc., is inexplicable without him. If Mr. Darwin believes that the events which he supposes to have occurred and the results we behold were undirected and undesigned, or if the physicist believes that the natural forces to which he refers phenomena are uncaused and undirected, no argument is needed to show that such belief is atheism. But the admission of the phenomena and of these natural processes and forces does not necessitate any such belief, nor even render it one whit less improbable than before.

Surely, too, the accidental element may play its part in Nature without negativing design in the the-

[1] In *American Journal of Science*, July, 1860, pp. 147–149.

ist's view. He believes that the earth's surface has been very gradually prepared for man and the existing animal races, that vegetable matter has through a long series of generations imparted fertility to the soil in order that it may support its present occupants, that even beds of coal have been stored up for man's benefit. Yet what is more accidental, and more simply the consequence of physical agencies, than the accumulation of vegetable matter in a peat-bog, and its transformation into coal? No scientific person at this day doubts that our solar system is a progressive development, whether in his conception he begins with molten masses, or aëriform or nebulous masses, or with a fluid revolving mass of vast extent, from which the specific existing worlds have been developed one by one. What theist doubts that the actual results of the development in the inorganic worlds are not merely compatible with design, but are in the truest sense designed results? Not Mr. Agassiz, certainly, who adopts a remarkable illustration of design directly founded on the nebular hypothesis, drawing from the position and times of the revolution of the world, so originated, "direct evidence that the physical world has been ordained in conformity with laws which obtain also among living beings." But the reader of the interesting exposition[1] will notice that the designed result has been brought to pass through what, speaking after the manner of men, might be called a chapter of accidents.

A natural corollary of this demonstration would

[1] In "Contributions to the Natural History of the United States," vol. i., pp. 128, 129.

seem to be, that a material connection between a series of created things—such as the development of one of them from another, or of all from a common stock—is highly compatible with their intellectual connection, namely, with their being designed and directed by one mind. Yet upon some ground which is not explained, and which we are unable to conjecture, Mr. Agassiz concludes to the contrary in the organic kingdoms, and insists that, *because* the members of such a series have an intellectual connection, "they cannot be the result of a material differentiation of the objects themselves,"[1] that is, they cannot have had a genealogical connection. But is there not as much intellectual connection between the successive generations of any species as there is between the several species of a genus, or the several genera of an order? As the intellectual connection here is realized through the material connection, why may it not be so in the case of species and genera? On all sides, therefore, the implication seems to be quite the other way.

Returning to the accidental element, it is evident that the strongest point against the compatibility of Darwin's hypothesis with design in Nature is made when natural selection is referred to as picking out those variations which are improvements from a vast number which are not improvements, but perhaps the contrary, and therefore useless or purposeless, and born to perish. But even here the difficulty is not peculiar; for Nature abounds with analogous instances. Some of our race are useless, or worse, as regards

[1] "Contributions to the Natural History of the United States," vol. i., p. 130; and *American Journal of Science*, July, 1860, p. 143.

the improvement of mankind; yet the race may be designed to improve, and may be actually improving. Or, to avoid the complication with free agency—the whole animate life of a country depends absolutely upon the vegetation, the vegetation upon the rain. The moisture is furnished by the ocean, is raised by the sun's heat from the ocean's surface, and is wafted inland by the winds. But what multitudes of rain-drops fall back into the ocean—are as much without a final cause as the incipient varieties which come to nothing! Does it therefore follow that the rains which are bestowed upon the soil with such rule and average regularity were not designed to support vegetable and animal life? Consider, likewise, the vast proportion of seeds and pollen, of ova and young—a thousand or more to one—which come to nothing, and are therefore purposeless in the same sense, and only in the same sense, as are Darwin's unimproved and unused slight variations. The world is full of such cases; and these must answer the argument—for we cannot, except by thus showing that it proves too much.

Finally, it is worth noticing that, though natural selection is scientifically explicable, variation is not. Thus far the cause of variation, or the reason why the offspring is sometimes unlike the parents, is just as mysterious as the reason why it is generally like the parents. It is now as inexplicable as any other origination; and, if ever explained, the explanation will only carry up the sequence of secondary causes one step farther, and bring us in face of a somewhat different problem, but which will have the same element of mystery that the problem of variation has now. Circumstances may preserve or may destroy the variations; man may use or direct them; but selection, whether artificial or natural, no more originates them than man originates the power which turns a wheel, when he dams a stream and lets the water fall upon it. The origination of this power is a question about efficient cause. The tendency of science in respect to this obviously is not toward the omnipotence of matter, as some suppose, but toward the omnipotence of spirit.

So the real question we come to is as to the way in which we are to conceive intelligent and efficient cause to be exerted, and upon what exerted. Are we bound to suppose efficient cause in all cases exerted upon nothing to evoke something into existence—and this thousands of times repeated, when a slight change in the details would make all the difference between successive species? Why may not the new species, or some of them, be designed diversifications of the old?

There are, perhaps, only three views of efficient cause which may claim to be both philosophical and theistic:

1. The view of its exertion at the beginning of time, endowing matter and created things with forces which do the work and produce the phenomena.

2. This same view, with the theory of insulated interpositions, or occasional direct action, engrafted upon it—the view that events and operations in general go on in virtue simply of forces communicated at the first, but that now and then, and only now and then, the Deity puts his hand directly to the work.

3. The theory of the immediate, orderly, and con-

stant, however infinitely diversified, action of the intelligent efficient Cause.

It must be allowed that, while the third is preëminently the Christian view, all three are philosophically compatible with design in Nature. The second is probably the popular conception. Perhaps most thoughtful people oscillate from the middle view toward the first or the third—adopting the first on some occasions, the third on others. Those philosophers who like and expect to settle all mooted questions will take one or the other extreme. The *Examiner* inclines toward, the *North American* reviewer fully adopts, the third view, to the logical extent of maintaining that "*the origin of an individual*, as well as the origin of a species or a genus, can be explained only by the *direct* action of an intelligent creative cause." To silence his critics, this is the line for Mr. Darwin to take; for it at once and completely relieves his scientific theory from every theological objection which his reviewers have urged against it.

At present we suspect that our author prefers the first conception, though he might contend that his hypothesis is compatible with either of the three. That it is also compatible with an atheistic or pantheistic conception of the universe, is an objection which, being shared by all physical, and some ethical or moral science, cannot specially be urged against Darwin's system. As he rejects spontaneous generation, and admits of intervention at the beginning of organic life, and probably in more than one instance, he is not wholly excluded from adopting the middle view, although the interventions he would allow are few and far back. Yet one interposition admits the principle as well as more. Interposition presupposes particular necessity or reason for it, and raises the question, when and how often it may have been necessary. It might be the natural supposition, if we had only one set of species to account for, or if the successive inhabitants of the earth had no other connections or resemblances than those which adaptation to similar conditions, which final causes in the narrower sense, might explain. But if this explanation of organic Nature requires one to "believe that, at innumerable periods in the earth's history, certain elemental atoms have been commanded suddenly to flash into living tissues," and this when the results are seen to be strictly connected and systematic, we cannot wonder that such interventions should at length be considered, not as interpositions or interferences, but rather—to use the reviewer's own language—as "exertions so frequent and beneficent that we come to regard them as the ordinary action of Him who laid the foundation of the earth, and without whom not a sparrow falleth to the ground."[1]

What does the difference between Mr. Darwin and his reviewer now amount to? If we say that according to one view the origination of species is *natural*, according to the other *miraculous*, Mr. Darwin agrees that "what is natural as much requires and presupposes an intelligent mind to render it so—that is, to effect it continually or at stated times—as what is supernatural does to effect it for once."[2] He merely

[1] *North American Review* for April, 1860, p. 506.

[2] *Vide* motto from Butler, prefixed to the second edition of Darwin's work.

inquires into the form of the miracle, may remind us that all recorded miracles (except the primal creation of matter) were transformations or actions in and upon natural things, and will ask how many times and how frequently may the origination of successive species be repeated before the supernatural merges in the natural.

In short, Darwin maintains that the origination of a species, no less than that of an individual, is natural; the reviewer, that the natural origination of an individual, no less than the origination of a species, requires and presupposes Divine power. *A fortiori*, then, the origination of a variety requires and presupposes Divine power. And so between the scientific hypothesis of the one and the philosophical conception of the other no contrariety remains. And so, concludes the *North American* reviewer, "a proper view of the nature of causation . . . . places the vital doctrine of the being and the providence of a God on ground that can never be shaken."[1] A worthy conclusion, and a sufficient answer to the denunciations and arguments of the rest of the article, so far as philosophy and natural theology are concerned. If a writer must needs use his own favorite dogma as a weapon with which to give *coup de grace* to a pernicious theory, he should be careful to seize his edge-tool by the handle, and not by the blade.

We can barely glance at a subsidiary philosophical objection of the *North American* reviewer, which the *Examiner* also raises, though less explicitly. Like all geologists, Mr. Darwin draws upon time in the

[1] *North American Review, loc. cit.*, p. 504.

most unlimited manner. He is not peculiar in this regard. Mr. Agassiz tells us that the conviction is "now universal, among well-informed naturalists, that this globe has been in existence for innumerable ages, and that the length of time elapsed since it first became inhabited cannot be counted in years;" Pictet, that the imagination refuses to calculate the immense number of years and of ages during which the faunas of thirty or more epochs have succeeded one another, and developed their long succession of generations. Now, the reviewer declares that such indefinite succession of ages is "virtually infinite," "lacks no characteristic of eternity except its name," at least, that "the difference between such a conception and that of the strictly infinite, if any, is not appreciable." But infinity belongs to metaphysics. Therefore, he concludes, Darwin supports his theory, not by scientific but by metaphysical evidence; his theory is "essentially and completely metaphysical in character, resting altogether upon that idea of 'the infinite' which the human mind can neither put aside nor comprehend." And so a theory which will be generally regarded as much too physical is transferred by a single syllogism to metaphysics.

Well, physical geology must go with it: for, even on the soberest view, it demands an indefinitely long time antecedent to the introduction of organic life upon our earth. *A fortiori* is physical astronomy a branch of metaphysics, demanding, as it does, still larger "instalments of infinity," as the reviewer calls them, both as to time and number. Moreover, far the

[1] *North American Review, loc. cit.*, p. 487, *et passim.*

greater part of physical inquiries now relate to molecular actions, which, a distinguished natural philosopher informs us, "we have to regard as the results of an *infinite* number of *infinitely* small material particles, acting on each other at *infinitely* small distances"—a triad of infinities—and so *physics* becomes the most *metaphysical* of sciences. Verily, if this style of reasoning is to prevail—

> "Thinking is but an idle waste of thought,
> And naught is everything, and everything is naught."

The leading objection of Mr. Agassiz is likewise of a philosophical character. It is, that species exist only "as categories of thought"—that, having no material existence, they can have had no material variation, and no material community of origin. Here the predication is of species in the subjective sense, the inference in the objective sense. Reduced to plain terms, the argument seems to be: Species are ideas; therefore the objects from which the idea is derived cannot vary or blend, and cannot have had a genealogical connection.

The common view of species is, that, although they are generalizations, yet they have a direct objective ground in Nature, which genera, orders, etc., have not. According to the succinct definition of Jussieu—and that of Linnæus is identical in meaning—a species is the perennial succession of similar individuals in continued generations. The species is the chain of which the individuals are the links. The sum of the genealogically-connected similar individuals constitutes the species, which thus has an actuality and ground of distinction not shared by genera and other groups which were not supposed to be genealogically connected. How a derivative hypothesis would modify this view, in assigning to species only a temporary fixity, is obvious. Yet, if naturalists adopt that hypothesis, they will still retain Jussieu's definition, which leaves untouched the question as to how and when the "perennial successions" were established. The practical question will only be, How much difference between two sets of individuals entitles them to rank under distinct species? and that is the practical question now, on whatever theory. The theoretical question is—as stated at the beginning of this article—whether these specific lines were always as distinct as now.

Mr. Agassiz has "lost no opportunity of urging the idea that, while species have no material existence, they yet exist as categories of thought in the same way [and only in the same way] as genera, families, orders, classes," etc. He

"has taken the ground that all the natural divisions in the animal kingdom are primarily distinct, founded upon different categories of characters, and that all exist in the same way, that is, as categories of thought, embodied in individual living forms. I have attempted to show that branches in the animal kingdom are founded upon different plans of structure, and for that very reason have embraced from the beginning representatives between which there could be no community of origin; that classes are founded upon different modes of execution of these plans, and therefore they also embrace representatives which could have no community of origin; that orders represent the different degrees of complication in the mode of execution of each class, and therefore embrace representatives which could not have a community of origin any more than the members of different classes or branches; that families are founded upon different patterns of form, and embrace representatives equally independent in their origin; that genera are founded upon ultimate peculiarities of structure, embracing representatives which, from the very nature of their peculiarities, could have no community of origin; and that, finally, species are based upon relations and proportions that exclude, as much as all the preceding distinctions, the idea of a common descent.

"As the community of characters among the beings belonging to these different categories arises from the intellectual connection which shows them to be categories of thought, they cannot be the result of a gradual material differentiation of the objects themselves. The argument on which these views are founded may be summed up in the following few words: Species, genera, families, etc., exist as thoughts, individuals as facts."[1]

An ingenious dilemma caps the argument:

"It seems to me that there is much confusion of ideas in the general statement of the variability of species so often repeated lately. If species do not exist at all, as the supporters of the transmutation theory maintain, how can they vary? and if individuals alone exist, how can the differences which may be observed among them prove the variability of species?"

Now, we imagine that Mr. Darwin need not be dangerously gored by either horn of this curious dilemma. Although we ourselves cherish old-fashioned prejudices in favor of the probable permanence, and therefore of a more stable objective ground of species, yet we agree—and Mr. Darwin will agree fully with Mr. Agassiz—that species, and he will add varieties, "exist as categories of thought," that is, as cognizable distinctions—which is all that we can make of the phrase here, whatever it may mean in the Aristotelian metaphysics. Admitting that species are only cate-

[1] In *American Journal of Science*, July, 1860, p. 143.

gories of thought, and not facts or things, how does this prevent the individuals, which are material things, from having varied in the course of time, so as to exemplify the present almost innumerable categories of thought, or embodiments of Divine thought in material forms, or—viewed on the human side—in forms marked with such orderly and graduated resemblances and differences as to suggest to our minds the idea of species, genera, orders, etc., and to our reason the inference of a Divine Original? We have no clear idea how Mr. Agassiz intends to answer this question, in saying that branches are founded upon different plans of structure, classes upon different mode of execution of these plans, orders on different degrees of complication in the mode of execution, families upon different patterns of form, genera upon ultimate peculiarities of structure, and species upon relations and proportions. That is, we do not perceive how these several "categories of thought" exclude the possibility or the probability that the individuals which manifest or suggest the thoughts had an ultimate community of origin.

Moreover, Mr. Darwin might insinuate that the particular philosophy of classification upon which this whole argument reposes is as purely hypothetical and as little accepted as is his own doctrine. If both are pure hypotheses, it is hardly fair or satisfactory to extinguish the one by the other. If there is no real contradiction between them, nothing is gained by the attempt.

As to the dilemma propounded, suppose we try it upon that category of thought which we call *chair*.

This is a genus, comprising a common chair (*Sella vulgaris*), arm or easy chair (*S. cathedra*), the rocking-chair (*S. oscillans*)—widely distributed in the United States—and some others, each of which has *sported*, as the gardeners say, into many varieties. But now, as the *genus* and the *species* have no material existence, how can they vary? If only individual chairs exist, how can the differences which may be observed among them prove the variability of the species? To which we reply by asking, Which does the question refer to, the category of thought, or the individual embodiment? If the former, then we would remark that our categories of thought vary from time to time in the readiest manner. And, although the Divine thoughts are eternal, yet they are manifested to us in time and succession, and by their manifestation only can we know them, how imperfectly! Allowing that what has no material existence can have had no material connection or variation, we should yet infer that what has intellectual existence and connection might have intellectual variation; and, turning to the individuals, which represent the species, we do not see how all this shows that they may not vary. Observation shows us that they do. Wherefore, taught by fact that successive individuals do vary, we safely infer that the idea must have varied, and that this variation of the individual representatives proves the variability of the species, whether objectively or subjectively regarded.

Each species or sort of chair, as we have said, has its varieties, and one species shades off by gradations into another. And—note it well—these numerous and successively slight variations and gradations, far from suggesting an accidental origin to chairs and to their forms, are very proofs of design.

Again, *edifice* is a generic category of thought. Egyptian, Grecian, Byzantine, and Gothic buildings are well-marked species, of which each individual building of the sort is a material embodiment. Now, the question is, whether these categories or ideas may not have been evolved, one from another in succession, or from some primal, less specialized, edificial category. What better evidence for such hypothesis could we have than the variations and grades which connect these species with each other? We might extend the parallel, and get some good illustrations of natural selection from the history of architecture, and the origin of the different styles under different climates and conditions. Two considerations may qualify or limit the comparison. One, that houses do not propagate, so as to produce continuing lines of each sort and variety; but this is of small moment on Agassiz's view, he holding that genealogical connection is not of the essence of a species at all. The other, that the formation and development of the ideas upon which human works proceed are gradual; or, as the same great naturalist well states it, "while human thought is consecutive, Divine thought is simultaneous." But we have no right to affirm this of Divine action.

We must close here. We meant to review some of the more general scientific objections which we thought not altogether tenable. But, after all, we are not so anxious just now to know whether the new theory is well founded on facts, as whether it would be harmless if it were. Besides, we feel quite unable to answer some of these objections, and it is pleasanter to take up those which one thinks he can.

Among the unanswerable, perhaps the weightiest of the objections, is that of the absence, in geological deposits, of vestiges of the intermediate forms which the theory requires to have existed. Here all that Mr. Darwin can do is to insist upon the extreme imperfection of the geological record and the uncertainty of negative evidence. But, withal, he allows the force of the objection almost as much as his opponents urge it—so much so, indeed, that two of his English critics turn the concession unfairly upon him, and charge him with actually basing his hypothesis upon these and similar difficulties—as if he held it because of the difficulties, and not in spite of them; a handsome return for his candor!

As to this imperfection of the geological record, perhaps we should get a fair and intelligible illustration of it by imagining the existing animals and plants of New England, with all their remains and products since the arrival of the Mayflower, to be annihilated; and that, in the coming time, the geologists of a new colony, dropped by the New Zealand fleet on its way to explore the ruins of London, undertake, after fifty years of examination, to reconstruct in a catalogue the flora and fauna of our day, that is, from the close of the glacial period to the present time. With all the advantages of a surface exploration, what a beggarly account it would be! How many of the land animals and plants which are enumerated in the Massachusetts official reports would it be likely to contain?

Another unanswerable question asked by the Boston reviewers is, Why, when structure and instinct or habit vary—as they must have varied, on Darwin's hypothesis—they vary together and harmoniously, instead of vaguely? We cannot tell, because we cannot tell why either varies at all. Yet, as they both do vary in successive generations—as is seen under domestication—and are correlated, we can only adduce the fact. Darwin may be precluded from our answer, but we may say that they vary together because designed to do so. A reviewer says that the chance of their varying together is inconceivably small; yet, if they do not, the variant individuals must all perish. Then it is well that it is not left to chance. To refer to a parallel case: before we were born, nourishment and the equivalent to respiration took place in a certain way. But the moment we were ushered into this breathing world, our actions promptly conformed, both as to respiration and nourishment, to the before unused structure and to the new surroundings.

"Now," says the *Examiner*, "suppose, for instance, the gills of an aquatic animal converted into lungs, while instinct still compelled a continuance under water, would not drowning ensue?" No doubt. But —simply contemplating the facts, instead of theorizing—we notice that young frogs do not keep their heads under water after ceasing to be tadpoles. The instinct promptly changes with the structure, without supernatural interposition—just as Darwin would have it, if the development of a variety or incipient species, though rare, were as natural as a metamorphosis.

"Or if a quadruped, not yet furnished with wings, were suddenly inspired with the instinct of a bird, and precipitated itself from a cliff, would not the descent be hazardously rapid?" Doubtless the animal would be no better supported than the objection. But Darwin makes very little indeed of voluntary efforts as a cause of change, and even poor Lamarck need not be caricatured. He never supposed that an elephant would take such a notion into his wise head, or that a squirrel would begin with other than short and easy leaps; yet might not the length of the leap be increased by practice?

The *North American* reviewer's position, that the higher brute animals have comparatively little instinct and no intelligence, is a heavy blow and great discouragement to dogs, horses, elephants, and monkeys. Thus stripped of their all, and left to shift for themselves as they may in this hard world, their pursuit and seeming attainment of knowledge under such peculiar difficulties are interesting to contemplate. However, we are not so sure as is the critic that instinct regularly increases downward and decreases upward in the scale of being. Now that the case of the bee is reduced to moderate proportions,[1] we know of nothing in instinct surpassing that of an animal so high as a bird, the talegal, the male of which plumes himself upon making a hot-bed in which to hatch his partner's eggs—which he tends and regulates the heat

[1] *Vide* article by Mr. C. Wright, in the *Mathematical Monthly* for May last.

of about as carefully and skillfully as the unplumed biped does an eccaleobion.[1]

As to the real intelligence of the higher brutes, it has been ably defended by a far more competent observer, Mr. Agassiz, to whose conclusions we yield a general assent, although we cannot quite place the best of dogs "in that respect upon a level with a considerable proportion of poor humanity," nor indulge the hope, or indeed the desire, of a renewed acquaintance with the whole animal kingdom in a future life.[2]

The assertion that acquired habitudes or instincts, and acquired structures, are not heritable, any breeder or good observer can refute.[3]

That "the human mind has become what it is out of a developed instinct,"[4] is a statement which Mr. Darwin nowhere makes, and, we presume, would not accept.[5] That he would have us believe that individ-

[1] *Vide Edinburgh Review* for January, 1860, article on "Acclimatization," etc.

[2] "Contributions, Essay on Classification," etc., vol. i., pp. 60–66.

[3] Still stronger assertions have recently been hazarded—even that heritability is of species only, not of individual characteristics—strangely overlooking the fundamental peculiarity of plants and animals, which is that they *reproduce*, and that the species is continued as such only because individuals reproduce their like.

. . . .

It has also been urged that variation is never cumulative. If this means that varieties are not capable of further variation, it is not borne out by observation. For cultivators and breeders well know that the main difficulty is to initiate a variation, and that new varieties are particularly prone to vary more.

[4] *North American Review*, April, 1860, p. 475.

[5] No doubt he would equally distinguish in kind between *instinct* (which physiologically is best conceived of as *congenital habit*, so that habits when inherited become instincts, just as varieties become fixed

ual animals acquire their instincts gradually,[1] is a statement which must have been penned in inadvertence both of the very definition of instinct, and of everything we know of in Mr. Darwin's book.

It has been attempted to destroy the very foundation of Darwin's hypothesis by denying that there are any wild varieties, to speak of, for natural selection to operate upon. We cannot gravely sit down to prove that wild varieties abound. We should think it just as necessary to prove that snow falls in winter. That variation among plants cannot be largely due to hy-

into races) and intelligence; but would maintain that both are endowments of the higher brutes and of man, however vastly and unequal their degree, and with whatever superaddition to simple intelligence in the latter.

[Prof. Joseph Le Conte, in POPULAR SCIENCE MONTHLY, September, 1875, refers to his definition of instinct as "inherited experience," published in April, 1871, as having been anticipated by that of Hering, as "inherited memory," in February of the same year. Doubtless the idea has been expressed by others long before us.]

To allow that "brutes have certain mental endowments in common with men," . . . . desires, affections, memory, simple imagination or the power of reproducing the sensible past in mental pictures, and even judgment of the simple or intuitive kind"—that "they compare and judge" ("Memoirs of American Academy," vol. viii., p. 118)—is to concede that the intellect of brutes really acts, so far as we know, like human intellect, as far it goes; for the philosophical logicians tell us all reasoning is reducible to a series of simple judgments. And Aristotle declares that even reminiscence—which is, we suppose, "reproducing the sensible past in mental pictures"—is a sort of reasoning (*τὸ ἀναμιμνήσκεσθαί ἐστιν οἷον συλλογισμός τις*).

On the other hand, Mr. Darwin's expectation that "psychology will be based on a new foundation, that of the *necessary* acquirement of each mental power and capacity by gradation," seems to come from a school of philosophy with which we have no sympathy.

[1] *American Journal of Science*, July, 1860, p. 146.

bridism, and that their variation in Nature is not essentially different from much that occurs in domestication, and, in the long-run, probably hardly less in amount, we could show if our space permitted.

As to the sterility of hybrids, that can no longer be insisted upon as absolutely true, nor be practically used as a test between species and varieties, unless we allow that hares and rabbits are of one species. That such sterility, whether total or partial, subserves a purpose in keeping species apart, and was so designed, we do not doubt. But the critics fail to perceive that this sterility proves nothing whatever against the derivative origin of the actual species; for it may as well have been intended to keep separate those forms which have reached a certain amount of divergence, as those which were always thus distinct.

The argument for the permanence of species, drawn from the identity with those now living of cats, birds, and other animals preserved in Egyptian catacombs, was good enough as used by Cuvier against St.-Hilaire, that is, against the supposition that time brings about a gradual alteration of whole species; but it goes for little against Darwin, unless it be proved that species never vary, or that the perpetuation of a variety necessitates the extinction of the parent breed. For Darwin clearly maintains—what the facts warrant—that the mass of a species remains fixed so long as it exists at all, though it may set off a variety now and then. The variety may finally supersede the parent form, or it may coexist with it; yet it does not in the least hinder the unvaried stock from continuing true to the breed, unless it crosses with it. The common law of inheritance may be expected to keep both the original and the variety mainly true as long as they last, and none the less so because they have given rise to occasional varieties. The tailless Manx cats, like the curtailed fox in the fable, have not induced the normal breeds to dispense with their tails, nor have the Dorkings (apparently known to Pliny) affected the permanence of the common sort of fowl.

As to the objection that the lower forms of life ought, on Darwin's theory, to have been long ago improved out of existence, and replaced by higher forms, the objectors forget what a vacuum that would leave below, and what a vast field there is to which a simple organization is best adapted, and where an advance would be no improvement, but the contrary. To accumulate the greatest amount of being upon a given space, and to provide as much enjoyment of life as can be under the conditions, is what Nature seems to aim at; and this is effected by diversification.

Finally, we advise nobody to accept Darwin's or any other derivative theory as true. The time has not come for that, and perhaps never will. We also advise against a simular credulity on the other side, in a blind faith that species—that the manifold sorts and forms of existing animals and vegetables—"have no secondary cause." The contrary is already not unlikely, and we suppose will hereafter become more and more probable. But we are confident that, if a derivative hypothesis ever is established, it will be so on a solid theistic ground.

Meanwhile an inevitable and legitimate hypothesis is on trial—an hypothesis thus far not untenable—a trial just now very useful to science, and, we conclude, not harmful to religion, unless injudicious assailants temporarily make it so.

One good effect is already manifest; its enabling the advocates of the hypothesis of a multiplicity of human species to perceive the double insecurity of their ground. When the races of men are admitted to be of one *species*, the corollary, that they are of one *origin*, may be expected to follow. Those who allow them to be of one species must admit an actual diversification into strongly-marked and persistent varieties, and so admit the basis of fact upon which the Darwinian hypothesis is built; while those, on the other hand, who recognize several or numerous human species, will hardly be able to maintain that such species were primordial and supernatural in the ordinary sense of the word.

The English mind is prone to positivism and kindred forms of materialistic philosophy, and we must expect the derivative theory to be taken up in that interest. We have no predilection for that school, but the contrary. If we had, we might have looked complacently upon a line of criticism which would indirectly, but effectively, play into the hands of positivists and materialistic atheists generally. The wiser and stronger ground to take is, that the derivative hypothesis leaves the argument for design, and therefore for a designer, as valid as it ever was; that to do any work by an instrument must require, and therefore presuppose, the exertion rather of more than of less power than to do it directly; that whoever would be a consistent theist should believe that Design in the natural world is coextensive with Providence, and hold as firmly to the one as he does to the other, in spite of the wholly similar and apparently insuperable difficulties which the mind encounters whenever it endeavors to develop the idea into a system, either in the material and organic, or in the moral world. It is enough, in the way of obviating objections, to show that the philosophical difficulties of the one are the same, and only the same, as of the other.

IV.

SPECIES AS TO VARIATION, GEOGRAPHICAL DISTRIBUTION, AND SUCCESSION.

(American Journal of Science and Arts, *May*, 1863.)

Étude sur l'Espèce, à l'Occasion d'une Révision de la Famille des Cupulifères, par M. Alphonse De Candolle.—This is the title of a paper by M. Alph. De Candolle, growing out of his study of the oaks. It was published in the November number of the *Bibliothèque Universelle*, and separately issued as a pamphlet. A less inspiring task could hardly be assigned to a botanist than the systematic elaboration of the genus *Quercus* and its allies. The vast materials assembled under De Candolle's hands, while disheartening for their bulk, offered small hope of novelty. The subject was both extremely trite and extremely difficult. Happily it occurred to De Candolle that an interest might be imparted to an onerous undertaking, and a work of necessity be turned to good account for science, by studying the oaks in view of the question of *species*.

What this term *species* means, or should mean, in natural history, what the limits of species, *inter se* or chronologically, or in geographical distribution, their modifications, actual or probable, their origin, and their destiny — these are questions which surge up from time to time; and now and then in the progress of science they come to assume a new and hopeful interest. Botany and zoölogy, geology, and what our author, feeling the want of a new term, proposes to name *epiontology*,[1] all lead up to and converge into this class of questions, while recent theories shape and point the discussion. So we look with eager interest to see what light the study of oaks by a very careful, experienced, and conservative botanist, particularly conversant with the geographical relations of plants, may throw upon the subject.

The course of investigation in this instance does not differ from that ordinarily pursued by working botanists; nor, indeed, are the theoretical conclusions other than those to which a similar study of other orders might not have equally led. The oaks afford a very good occasion for the discussion of questions which press upon our attention, and perhaps they offer peculiarly good materials on account of the number of fossil species.

Preconceived notions about species being laid aside, thé specimens in hand were distributed, accord-

[1] A name which, at the close of his article, De Candolle proposes for *the study of the succession of organized beings*, to comprehend, therefore, palæontology and all included under what is called geographical botany and zoölogy—the whole forming a science parallel to geology—the latter devoted to the history of unorganized bodies, the former, to that of organized beings, as respects origin, distribution, and succession. We are not satisfied with the word, notwithstanding the precedent of *palæontology;* since *ontology*, the science of being, has an established meaning as referring to mental existence—i. e., is a synonym or a department of metaphysics.

ing to their obvious resemblances, into groups of apparently identical or nearly identical forms, which were severally examined and compared. Where specimens were few, as from countries little explored, the work was easy, but the conclusions, as will be seen, of small value. The fewer the materials, the smaller the likelihood of forms intermediate between any two, and—what does not appear being treated upon the old law-maxim as non-existent—species are readily enough defined. Where, however, specimens abound, as in the case of the oaks of Europe, of the Orient, and of the United States, of which the specimens amounted to hundreds, collected at different ages, in varied localities, by botanists of all sorts of views and predilections—here alone were data fit to draw useful conclusions from. Here, as De Candolle remarks, he had every advantage, being furnished with materials more complete than any one person could have procured from his own herborizations, more varied than if he had observed a hundred times over the same forms in the same district, and more impartial than if they had all been amassed by one person with his own ideas or predispositions. So that vast herbaria, into which contributions from every source have flowed for years, furnish the best possible data—at least are far better than any practicable amount of personal herborization —for the comparative study of related forms occurring over wide tracts of territory. But as the materials increase, so do the difficulties. Forms, which appeared totally distinct, approach or blend through intermediate gradations; characters, stable in a limited number of instances or in a limited district, prove unstable occasionally, or when observed over a wider area; and the practical question is forced upon the investigator, What here is probably fixed and specific, and what is variant, pertaining to individual, variety, or race?

In the examination of these rich materials, certain characters were found to vary upon the same branch, or upon the same tree, sometimes according to age or development, sometimes irrespective of such relations or of any assignable reasons. Such characters, of course, are not specific, although many of them are such as would have been expected to be constant in the same species, and are such as generally enter into specific definitions. Variations of this sort, De Candolle, with his usual painstaking, classifies and tabulates, and even expresses numerically their frequency in certain species. The results are brought well to view in a systematic enumeration:

1. Of characters which *frequently* vary upon the same branch: over a dozen such are mentioned.

2. Of those which *sometimes* vary upon the same branch: a smaller number of these are mentioned.

3. Those so rare that they might be called monstrosities.

Then he enumerates characters, ten in number, which he has never found to vary on the same branch, and which, therefore, may better claim to be employed as specific. But, as among them he includes the duration of the leaves, the size of the cupule, and the form and size of its scales, which are by no means quite uniform in different trees of the same species, even these characters must be taken with allowance. In fact, having first brought together, as groups of the lowest order, those forms which varied upon the same stock, he next had to combine similarly various forms which, though not found associated upon the same branch, were thoroughly blended by intermediate degrees:

"The lower groups (varieties or races) being thus constituted, I have given the rank of *species* to the groups next above these, which differ in other respects, i. e., either in characters which were not found united upon certain individuals, or in those which do not show transitions from one individual to another. For the oaks of regions sufficiently known, the species thus formed rest upon satisfactory bases, of which the proof can be furnished. It is quite otherwise with those which are represented in our herbaria by single or few specimens. These are *provisional species*—species which may hereafter fall to the rank of simple varieties. I have not been inclined to prejudge such questions; indeed, in this regard, I am not disposed to follow those authors whose tendency is, as they say, to reunite species. I never reunite them without proof in each particular case; while the botanists to whom I refer do so on the ground of analogous variations or transitions occurring in the same genus or in the same family. For example resting on the fact that *Quercus Ilex*, *Q. coccifera*, *Q. acutifolia*, etc., have the leaves sometimes entire and sometimes toothed upon the same branch, or present transitions from one tree to another, I might readily have united my *Q. Tlapuxahuensis* to *Q. Sartorii* of Liebmann, since these two differ only in their entire or their toothed leaves. From the fact that the length of the peduncle varies in *Q. Robur* and many other oaks, I might have combined *Q. Seemannii* Liebm. with *Q. salicifolia* Née. I have not admitted these inductions, but have demanded visible proof in each particular case. Many species are thus left as provisional; but, in proceeding thus, the progress of the science will be more regular, and the synonymy less dependent upon the caprice or the theoretical opinions of each author."

This is safe and to a certain degree judicious, no doubt, as respects published species. Once admitted, they may stand until they are put down by evidence, direct or circumstantial. Doubtless a species may rightfully be condemned on good circumstantial evidence. But what course does De Candolle pursue in the case—of every-day occurrence to most working botanists, having to elaborate collections from countries not so well explored as Europe—when the forms in question, or one of the two, are as yet unnamed? Does he introduce as a new species every form which he cannot connect by ocular proof with a near relative, from which it differs only in particulars which he sees are inconstant in better known species of the same group? We suppose not. But, if he does, little improvement for the future upon the state of things revealed in the following quotation can be expected:

"In the actual state of our knowledge, after having seen nearly all the original specimens, and in some species as many as two hundred representatives from different localities, I estimate that, out of the three hundred species of *Cupuliferæ* which will be enumerated in the Prodromus, two-thirds at least are *provisional* species. In general, when we consider what a multitude of species were described from a single specimen, or from the forms of a single locality, of a single country, or are badly described, it is difficult to believe that above one-third of the actual species in botanical works will remain unchanged."

Such being the results of the *want* of adequate knowledge, how is it likely to be when our knowledge is largely increased? The judgment of so practised a botanist as De Candolle is important in this regard, and it accords with that of other botanists of equal experience.

"They are mistaken," he pointedly asserts, "who repeat that the greater part of our species are clearly limited, and that the doubtful species are in a feeble minority. This seemed to be true, so long as a genus was imperfectly known, and its species were founded upon few specimens, that is to say, were provisional. Just as we come to know them better, intermediate forms flow in, and doubts as to specific limits augment."

De Candolle insists, indeed, in this connection, that the higher the rank of the groups, the more definite their limitation, or, in other terms, the fewer the ambiguous or doubtful forms; that genera are more strictly limited than species, tribes than genera, orders than tribes, etc. We are not convinced of this. Often where it has appeared to be so, advancing discovery has brought intermediate forms to light, perplexing to the systematist. "They are mistaken," we think more than one systematic botanist will say, "who repeat that the greater part of our *natural orders and tribes* are absolutely limited," however we may agree that we will limit them. Provisional genera we suppose are proportionally hardly less common than provisional species; and hundreds of genera are kept up on considerations of general propriety or general convenience, although well known to shade off into adjacent ones by complete gradations. Somewhat of this greater fixity of higher groups, therefore, is rather apparent than real. On the other hand, that varieties should be less definite than species, follows from the very terms employed. They are ranked as varieties, rather than species, just because of their less definiteness.

Singular as it may appear, we have heard it denied that spontaneous varieties occur. De Candolle makes the important announcement that, in the oak genus, the best known species are just those which present the greatest number of spontaneous varieties and sub-varieties. The maximum is found in *Q. Robur*, with twenty-eight varieties, all spontaneous. Of *Q. Lusitanica* eleven varieties are enumerated, of *Q. Calliprinos* ten, of *Q. coccifera* eight, etc. And he significantly adds that "these very species which offer such numerous modifications are themselves ordinarily surrounded by other forms, provisionally called species, because of the absence of known transitions or variations, but to which some of these will probably have to be joined hereafter." The inference is natural, if not inevitable, that the difference between such species and such varieties is only one of degree, either as to amount of divergence, or of hereditary fixity, or as to the frequency or rarity at the present time of intermediate forms.

This brings us to the second section of De Candolle's article, in which he passes on, from the observation of the present forms and affinities of cupuliferous plants, to the consideration of their probable history and origin. Suffice it to say, that he frankly accepts the inferences derived from the whole course of observation, and contemplates a probable historical connection between congeneric species. He accepts and, by various considerations drawn from the geographical distribution of European *Cupuliferæ*, fortifies the conclusion—long ago arrived at by Edward Forbes—that the present species, and even some of their varieties, date back to about the close of the Tertiary epoch, since which time they have been subject to frequent and great changes of habitation or limitation, but without appreciable change of specific form or character; that is, without profounder changes than those within which a species at the present time is known to vary. Moreover, he is careful to state that he is far from concluding that the time of the appearance of a species in Europe at all indicates the time of its origin. Looking back still further into the Tertiary epoch, of which the vegetable remains indicate many analogous, but few, if any, identical forms, he concludes, with Heer and others, that specific changes of form, as well as changes of station, are to be presumed; and, finally, that "the theory of a succession of forms through the deviation of anterior forms is the most natural hypothesis, and the most accordant with the known facts in palæontology, geographical botany and zoölogy, of anatomical structure and classification: but direct proof of it is wanting, and moreover, if true, it must have taken place very slowly; so slowly, indeed, that its effects are discernible only after a lapse of time far longer than our historic epoch."

In contemplating the present state of the species of *Cupuliferæ* in Europe, De Candolle comes to the conclusion that, while the beech is increasing, and extending its limits southward and westward (at the expense of *Coniferæ* and birches), the common oak, to some extent, and the Turkey oak decidedly, are diminishing and retreating, and this wholly irrespective of man's agency. This is inferred of the Turkey oak from the great gaps found in its present geographical area, which are otherwise inexplicable, and which he regards as plain indications of a partial extinction.

Community of descent of all the individuals of species is of course implied in these and all similar reasonings.

An obvious result of such partial extinction is clearly enough brought to view. The European oaks (like the American species) greatly tend to vary; that is, they manifest an active disposition to produce new forms. Every form tends to become hereditary, and so to pass from the state of mere variation to that of race; and of these competing incipient races some only will survive. *Quercus Robur* offers a familiar illustration of the manner in which one form may in the course of time become separated into two or more distinct ones.

To Linnæus this common oak of Europe was all of one species. But of late years the greater number of European botanists have regarded it as including three species, *Q. pedunculata*, *Q. sessiliflora*, and *Q. pubescens.* De Candolle looks with satisfaction to the independent conclusion which he reached from a long and patient study of the forms (and which Webb, Gay, Bentham, and others, had equally reached), that the view of Linnæus was correct, inasmuch as it goes to show that the idea and the practical application of the term *species* have remained unchanged during the century which has elapsed since the publication of the "Species Plantarum." But, the idea remaining unchanged, the facts might appear under a different aspect, and the conclusion be different, under a slight and very supposable change of circumstances. Of the twenty-eight spontaneous varieties of *Q. Robur*, which De Candolle recognizes, all but six, he remarks, fall naturally under the three sub-species, *pedunculata*, *sessiliflora*, and *pubescens*, and are therefore forms grouped around these as centres; and, moreover, the few connecting forms are by no means the most common. Were these to die out, it is clear that the three forms which have already been so frequently taken for species would be what the group of four or five provisionally admitted species which closely surround *Q. Robur* now are. The best example of such a case, as having in all probability occurred through geographical segregation and partial extinction, is that of the cedar, thus separated into the Deodar, the Lebanon, and the Atlantic cedars—a case admirably worked out by Dr. Hooker two or three years ago.[1]

A special advantage of the *Cupuliferæ* for determining the probable antiquity of existing species in Europe, De Candolle finds in the size and character of their fruits. However it may be with other plants (and he comes to the conclusion generally that marine currents and all other means of distant transport have played only a very small part in the actual dispersion of species), the transport of acorns and chestnuts by natural causes across an arm of the sea in a condition to germinate, and much more the spontaneous establishment of a forest of oaks or chestnuts in this way, De Candolle conceives to be fairly impossible in itself, and contrary to all experience. From such considerations, i. e., from the actual dispersion of the existing species (with occasional aid from post-tertiary deposits), it is thought to be shown that the principal *Cupuliferæ* of the Old World attained their actual extension

[1] *Natural History Review*, January, 1862.

before the present separation of Sicily, Sardinia and Corsica, and of Britain, from the European Continent.

This view once adopted, and this course once entered upon, has to be pursued farther. *Quercus Robur* of Europe with its bevy of admitted derivatives, and its attending species only provisionally admitted to that rank, is very closely related to certain species of Eastern Asia, and of Oregon and California —so closely that "a view of the specimens by no means forbids the idea that they have all originated from *Q. Robur*, or have originated, with the latter, from one or more preceding forms so like the present ones that a naturalist could hardly know whether to call them species or varieties." Moreover, there are fossil leaves from diluvian deposits in Italy, figured by Gaudin, which are hardly distinguishable from those of *Q. Robur* on the one hand, and from those of *Q. Douglasii*, etc., of California, on the other. No such leaves are found in any tertiary deposit in Europe; but such are found of that age, it appears, in Northwest America, where their remote descendants still flourish. So that the probable genealogy of *Q. Robur*, traceable in Europe up to the commencement of the present epoch, looks eastward and far into the past on far-distant shores.

Quercus Ilex, the evergreen oak of Southern Europe and Northern Africa, reveals a similar archæology; but its presence in Algeria leads De Candolle to regard it as a much more ancient denizen of Europe than *Q. Robur*; and a Tertiary oak, *Q. ilicoides*, from a very old Miocene bed in Switzerland, is thought to be one of its ancestral forms. This high antiquity once established, it follows almost of course that the very nearly-related species in Central Asia, in Japan, in California, and even our own live-oak with its Mexican relatives, may probably enough be regarded as early offshoots from the same stock with *Q. Ilex.*

In brief—not to continue these abstracts and remarks, and without reference to Darwin's particular theory (which De Candolle at the close very fairly considers)—if existing species, or many of them, are as ancient as they are now generally thought to be, and were subject to the physical and geographical changes (among them the coming and the going of the glacial epoch) which this antiquity implies; if in former times they were as liable to variation as they now are; and if the individuals of the same species may claim a common local origin, then we cannot wonder that "the theory of a succession of forms by deviations of anterior forms" should be regarded as "the most natural hypothesis," nor at the general advance made toward its acceptance.

The question being, not, how plants and animals originated, but, how came the existing animals and plants to be just where they are and what they are, it is plain that naturalists interested in such inquiries are mostly looking for the answer in one direction. The general drift of opinion, or at least of expectation, is exemplified by this essay of De Candolle; and the set and force of the current are seen by noticing how it carries along naturalists of widely different views and prepossessions—some faster and farther than others—but all in one way. The tendency is, we may say, to extend the law of continuity, or something analogous to it, from inorganic to organic Nature, and in the latter to connect the present with the past in some sort of material connection. The generalization may indeed be expressed so as not to assert that the connection is genetic, as in Mr. Wallace's formula: "Every species has come into existence coincident both in time and space with preëxisting closely-allied species." Edward Forbes, who may be called the originator of this whole line of inquiry, long ago expressed a similar view. But the only material sequence we know, or can clearly conceive, in plants and animals, is that from parent to progeny; and, as De Candolle implies, the origin of species and that of races can hardly be much unlike, nor governed by other than the same laws, whatever these may be.

The progress of opinion upon this subject in one generation is not badly represented by that of De Candolle himself, who is by no means prone to adopt new views without much consideration. In an elementary treatise published in the year 1835, he adopted and, if we rightly remember, vigorously maintained, Schouw's idea of the double or multiple origin of species, at least of some species—a view which has been carried out to its ultimate development only perhaps by Agassiz, in the denial of any necessary genetic connection among the individuals of the same species, or of any original localization more restricted than the area now occupied by the species. But in 1855, in his "Géographie Botanique," the multiple hypothesis, although in principle not abandoned, loses its point, in view of the probable high antiquity of existing species. The actual vegetation of the world being now regarded as a continuation, through numerous geological, geographical, and more recently historical changes, of anterior vegetations, the actual distribution of plants is seen to be a consequence of preceding conditions; and geological considerations, and these alone, may be expected to explain all the facts—many of them so curious and extraordinary—of the actual geographical distribution of the species. In the present essay, not only the distribution but the origin of congeneric species is regarded as something derivative; whether derived by slow and very gradual changes in the course of ages, according to Darwin, or by a sudden, inexplicable change of their tertiary ancestors, as conceived by Heer, De Candolle hazards no opinion. It may, however, be inferred that he looks upon "natural selection" as a real, but insufficient cause; while some curious remarks upon the number of monstrosities annually produced, and the possibility of their enduring, may be regarded as favorable to Heer's view.

As an index to the progress of opinion in the direction referred to, it will be interesting to compare Sir Charles Lyell's well-known chapters of twenty or thirty years ago, in which the permanence of species was ably maintained, with his treatment of the same subject in a work just issued in England, which, however, has not yet reached us.

A belief of the derivation of species may be maintained along with a conviction of great persistence of specific characters. This is the idea of the excellent Swiss vegetable palæontologist, Heer, who imagines a sudden change of specific type at certain periods, and perhaps is that of Pictet. Falconer adheres to somewhat similar views in his elaborate paper on elephants, living and fossil, in the *Natural History Review* for January last. Noting that "there is clear evidence of the true mammoth having existed in America long after the period of the northern drift, when the surface of the country had settled down into its present form, and also in Europe so late as to have been a contemporary of the Irish elk, and on the other hand that it existed in England so far back as before the deposition of the bowlder clay; also that four well-defined species of fossil elephant are known to have existed in Europe; that "a vast number of the remains of three of these species have been exhumed over a large area in Europe; and, even in the geological sense, an enormous interval of time has elapsed between the formation of the most ancient and the most recent of these deposits, quite sufficient to test the persistence of specific characters in an elephant," he presents the question, "Do, then, the successive elephants occurring in these strata show any signs of a passage from the older form into the newer?"

To which the reply is: "If there is one fact which is impressed on the conviction of the observer with more force than any other, it is the persistence and uniformity of the characters of the molar teeth in the earliest known mammoth and his most modern successor. . . . Assuming the observation to be correct, what strong proof does it not afford of the persistence and constancy, throughout vast intervals of time, of the distinctive characters of those organs which are most concerned in the existence and habits of the species? If we cast a glance back on the long vista of physical changes which our planet has undergone since the Neozoic epoch, we can nowhere detect signs of a revolution more sudden and pronounced, or more important in its results, than the intercalation and sudden disappearance of the glacial period. Yet the 'dicyclotherian' mammoth lived before it, and passed through the ordeal of all the hard extremities it involved, bearing his organs of locomotion and digestion all but unchanged. Taking the group of four European fossil species above enumerated, do they show any signs in the successive deposits of a transition from the one form into the other? Here again the result of my observation, in so far as it has extended over the European area, is, that the specific characters of the molars are constant in each, within a moderate range of variation, and that we nowhere meet with intermediate forms." . . . . Dr. Falconer continues (page 80):

"The inferences which I draw from these facts are not opposed to one of the leading propositions of Darwin's theory. With him, I have no faith in the opinion that the mammoth and other extinct elephants made their appearance suddenly, after the type in which their fossil remains are presented to us. The most rational view seems to be, that they are in some shape the modified descendants of earlier progenitors. But if the asserted facts be correct, they seem clearly to indicate that the older elephants of Europe, such as *E. meridionalis* and *E. antiquus*, were not the stocks from which the later species, *E. primigenius* and *E. Africanus* sprung, and that we must look elsewhere for their origin. The nearest affinity, and that a very close one, of the European *E. meridionalis* is with the Miocene *E. planifrons* of India; and of *E. primigenius*, with the existing India species.

"Another reflection is equally strong in my mind—that the

means which have been adduced to explain the origin of the species by 'natural selection,' or a process of variation from external influences, are inadequate to account for the phenomena. The law of phyllotaxis, which governs the evolution of leaves around the axis of a plant, is as nearly constant in its manifestation as any of the physical laws connected with the material world. Each instance, however different from another, can be shown to be a term of some series of continued fractions. When this is coupled with the geometrical law governing the evolution of form, so manifest in some departments of the animal kingdom, e. g., the spiral shells of the Mollusca, it is difficult to believe that there is not, in Nature, a deeper-seated and innate principle, to the operation of which natural selection is merely an adjunct. The whole range of the Mammalia, fossil and recent, cannot furnish a species which has had a wider geographical distribution, and passed through a longer term of time, and through more extreme changes of climatal conditions, than the mammoth. If species are so unstable, and so susceptible of mutation through such influences, why does that extinct form stand out so signally a monument of stability? By his admirable researches and earnest writings, Darwin has, beyond all his contemporaries, given an impulse to the philosophical investigation of the most backward and obscure branch of the biological sciences of his day; he has laid the foundations of a great edifice; but he need not be surprised if, in the progress of erection, the superstructure is altered by his successors, like the Duomo of Milan from the Roman to a different style of architecture."

Entertaining ourselves the opinion that something more than natural selection is requisite to account for the orderly production and succession of species, we offer two incidental remarks upon the above extract.

1. We find in it—in the phrase "natural selection, or a process of variation from external influences"—an example of the very common confusion of two distinct things, viz., *variation* and *natural*

selection. The former has never yet been shown to have its cause in "external influences," nor to occur at random. As we have elsewhere insisted, if not inexplicable, it has never been explained; all we can yet say is, that plants and animals are prone to vary, and that some conditions favor variation. Perhaps in this Dr. Falconer may yet find what he seeks: for "it is difficult to believe that there is not in [its] nature a deeper-seated and innate principle, to the operation of which natural selection is merely an adjunct." The latter, which is the *ensemble* of the external influences, including the competition of the individuals themselves, picks out certain variations as they arise, but in no proper sense can be said to originate them.

2. Although we are not quite sure how Dr. Falconer intends to apply the law of phyllotaxis to illustrate his idea, we fancy that a pertinent illustration may be drawn from it, in this way. There are two *species* of phyllotaxis, perfectly distinct, and, we suppose, not mathematically reducible the one to the other, viz.: (1.) That of alternate leaves, with its varieties; and (2.) That of verticillate leaves, of which opposite leaves present the simplest case. That, although generally constant, a change from one variety of alternate phyllotaxis to another should occur on the same axis, or on successive axes, is not surprising, the different sorts being terms of a regular series—although, indeed, we have not the least idea as to how the change from the one to the other comes to pass. But it is interesting, and in this connection perhaps instructive, to remark that, while some dicotyledonous plants hold to the verticillate, i. e., opposite-leaved phyllotaxis

throughout, a larger number—through the operation of some deep-seated and innate principle, which we cannot fathom—change abruptly into the other species at the second or third node, and change back again in the flower, or else effect a synthesis of the two species in a manner which is puzzling to understand. Here is a change from one fixed law to another, as unaccountable, if not as great, as from one specific form to another.

An elaborate paper on the vegetation of the Tertiary period in the southeast of France, by Count Gaston de Saporta, published in the *Annales des Sciences Naturelles* in 1862, vol. xvi., pp. 309–344—which we have not space to analyze—is worthy of attention from the general inquirer, on account of its analysis of the Tertiary flora into its separate types, Cretaceous, Austral, Tropical, and Boreal, each of which has its separate and different history—and for the announcement that "the *hiatus*, which, in the idea of most geologists, intervened between the close of the Cretaceous and the beginning of the Tertiary, appears to have had no existence, so far as concerns the vegetation; that in general it was not by means of a total overthrow, followed by a complete new emission of species, that the flora has been renewed at each successive period; and that while the plants of Southern Europe inherited from the Cretaceous period more or less rapidly disappeared, as also the austral forms, and later the tropical types (except the laurel, the myrtle, and the *Chamærops humilis*), the boreal types, coming later, survived all the others, and now compose, either in Europe, or in the north of Asia, or in North America, the basis of the actual arborescent vegetation. Especially "a very considerable number of forms nearly identical with tertiary forms now exist in America, where they have found, more easily than in our [European] soil—less vast and less extended southward—refuge from ulterior revolutions." The extinction of species is attributed to two kinds of causes; the one material or physical, whether slow or rapid; the other inherent in the nature of organic beings, incessant, but slow, in a manner latent, but somehow assigning to the species, as to the individuals, a limited period of existence, and, in some equally mysterious but wholly natural way, connected with the development of organic types: "By *type* meaning a collection of vegetable forms constructed upon the same plan of organization, of which they reproduce the essential lineaments with certain secondary modifications, and which appear to run back to a common point of departure."

In this community of types, no less than in the community of certain existing species, Saporta recognizes a prolonged material union between North America and Europe in former times. Most naturalists and geologists reason in the same way—some more cautiously than others—yet perhaps most of them seem not to perceive how far such inferences imply the doctrine of the common origin of related species.

For obvious reasons such doctrines are likely to find more favor with botanists than with zoölogists. But with both the advance in this direction is seen to have been rapid and great; yet to us not unexpected. We note, also, an evident disposition, notwithstanding some endeavors to the contrary, to allow derivative hypotheses to stand or fall upon their own merits—to have indeed upon philosophical grounds certain presumptions in their favor—and to be, perhaps, quite as capable of being turned to good account as to bad account in natural theology.[1]

Among the leading naturalists, indeed, such views —taken in the widest sense—have one and, so far as we are now aware, only one thoroughgoing and thoroughly consistent opponent, viz., Mr. Agassiz.

Most naturalists take into their very conception of a species, explicitly or by implication, the notion of a material connection resulting from the descent of the individuals composing it from a common stock, of local origin. Agassiz wholly eliminates community of descent from his idea of species, and even conceives a species to have been as numerous in individuals and as wide-spread over space, or as segregated in discontinuous spaces, from the first as at the later period.

The station which it inhabits, therefore, is with

[1] What the Rev. Principal Tulloch remarks in respect to the philosophy of miracles has a pertinent application here. We quote at second hand:

"The stoutest advocates of interference can mean nothing more than that the Supreme Will has so moved the hidden springs of Nature that a new issue arises on given circumstances. The ordinary issue is supplanted by a higher issue. The essential facts before us are a certain set of phenomena, and a Higher Will moving them. How moving them? is a question for human definition; the answer to which does not and cannot affect the divine meaning of the change. Yet when we reflect that this Higher Will is everywhere reason and wisdom, it seems a juster as well as a more comprehensive view to regard it as operating by subordination and evolution, rather than by interference or violation."

other naturalists in no wise essential to the species, and may not have been the region of its origin. In Agassiz's view the habitat is supposed to mark the origin, and to be a part of the character of the species. The habitat is not merely the place where it is, but a part of what it is.

Most naturalists recognize varieties of species; and many, like De Candolle, have come to conclude that varieties of the highest grade, or races, so far partake of the characteristics of species, and are so far governed by the same laws, that it is often very difficult to draw a clear and certain distinction between the two. Agassiz will not allow that varieties or races exist in Nature, apart from man's agency.

Most naturalists believe that the origin of species is supernatural, their dispersion or particular geographical area, natural, and their extinction, when they disappear, also the result of physical causes. In the view of Agassiz, if rightly understood, all three are equally independent of physical cause and effect, are equally supernatural.

In comparing preceding periods with the present and with each other, most naturalists and palæontologists now appear to recognize a certain number of species as having survived from one epoch to the next, or even through more than one formation, especially from the Tertiary into the post-Tertiary period, and from that to the present age. Agassiz is understood to believe in total extinctions and total new creations at each successive epoch, and even to recognize no existing species as ever contemporary with extinct ones, except in the case of recent exterminations.

These peculiar views, if sustained, will effectually dispose of every form of derivative hypothesis.

Returning for a moment to De Candolle's article, we are disposed to notice his criticism of Linnæus's "definition" of the term *species* (*Philosophia Botanica*, No. 157): "*Species tot numeramus quot diversæ formæ in principio sunt creatæ*"—which he declares illogical, inapplicable, and the worst that has been propounded. "So, to determine if a form is specific, it is necessary to go back to its origin, which is impossible. A definition by a character which can never be verified is no definition at all."

Now, as Linnæus practically applied the idea of species with a sagacity which has never been surpassed, and rarely equaled, and indeed may be said to have fixed its received meaning in natural history, it may well be inferred that in the phrase above cited he did not so much undertake to frame a logical *definition*, as to set forth the *idea* which, in his opinion, lay at the foundation of species; on which basis A. L. Jussieu did construct a logical definition—"Nunc rectius definitur perennis individuorum similium successio continuata generatione renascentiun." The fundamental idea of species, we would still maintain, is that of a chain of which genetically-connected individuals are the links. That, in the practical recognition of species, the essential characteristic has to be *inferred*, is no great objection—the general fact that like engenders like being an induction from a vast number of instances, and the only assumption being that of the uniformity of Nature. The idea of gravitation, that of the atomic constitution of matter, and the like, equally have to be verified inferentially. If we still hold to the idea of Linnæus, and of Agassiz, that existing species were created independently and essentially all at once at the beginning of the present era, we could not better the propositions of Linnæus and of Jussieu. If, on the other hand, the time has come in which we may accept, with De Candolle, their successive origination, at the commencement of the present era or before, and even by derivation from other forms, then the "*in principio*" of Linnæus will refer to that time, whenever it was, and his proposition be as sound and wise as ever.

In his "Géographie Botanique" (ii., 1068–1077) De Candolle discusses this subject at length, and in the same interest. Remarking that of the two great facts of species, viz., *likeness among the individuals*, and *genealogical connection*, zoölogists have generally preferred the latter,[1] while botanists have been divided in opinion, he pronounces for the former as the essential thing, in the following argumentative statement:

"Quant à moi, j'ai été conduit, dans ma definition de l'espèce, à mettre décidément la ressemblance au-dessus de caractères de succession. Ce n'est pas seulement à cause des circonstances propres au règne végétal, dont je m'occupe exclusivement; ce n'est pas non plus afin de sortir ma définition des théories et de la rendre le plus possible utile aux naturalistes descripteurs et nomenclateurs, c'est aussi par un motif philosophique. En toute chose il faut aller au fond des questions, quand on le peut. Or, pourquoi la reproduction est-elle possible, habituelle, féconde indefiniment, entre des êtres organisés que nous dirons de la

[1] Particularly citing Flourens: "La ressemblance n'est qu'une condition secondaire; la condition essentielle est la descendance: ce n'est pas la ressemblance, c'est la succession des individus, qui fait l'espèce."

même espèce? Parce qu'ils se ressemblent et uniquement à cause de cela. Lorsque deux espèces ne peuvent, ou, s'il s'agit d'animaux supérieurs, ne peuvent et ne veulent se croiser, c'est qu'elles sont très differentes. Si l'on obtient des croisements, c'est que les individus sont analogues; si ces croisements donnent des produits féconds, c'est que les individus étaient plus analogues; si ces produits eux-mêmes sont féconds, c'est que la ressemblance était plus grande; s'ils sont fécond habituellement et indéfiniment, c'est que la ressemblance intérieure et extérieure était très grande. Ainsi le degré de ressemblance est le fond; la reproduction en est seulement la manifestation et la mesure, et il est logique de placer la cause au-dessus de l'effet."

We are not yet convinced. We still hold that genealogical connection, rather than mutual resemblance, is the fundamental thing—first on the ground of fact, and then from the philosophy of the case. Practically, no botanist can say what amount of dissimilarity is compatible with unity of species; in wild plants it is sometimes very great, in cultivated races often enormous. De Candolle himself informs us that the different variations which the same oak-tree exhibits are significant indications of a disposition to set up separate varieties, which becoming hereditary may constitute a race; he evidently looks upon the extreme forms, say of *Quercus Robur*, as having thus originated; and on this ground, inferred from transitional forms, and not from their mutual resemblance, he includes them in that species. This will be more apparent should the discovery of transitions, which he leads us to expect, hereafter cause the four provisional species which attend *Q. Robur* to be merged in that species. It may rightly be replied that this conclusion would be arrived at from the likeness step by step in the series of forms; but the cause of the likeness here is obvious. And this brings in our "*motif philosophique.*"

Not to insist that the likeness is after all the variable, not the constant, element—to learn which is the essential thing, resemblance among individuals or their genetic connection—we have only to ask which can be the cause of the other.

In hermaphrodite plants (the normal case), and even as the question is ingeniously put by De Candolle in the above extract, the former surely cannot be the *cause* of the latter, though it may, in case of crossing, offer *occasion*. But, on the ground of the most fundamental of all things in the constitution of plants and animals—the fact incapable of further analysis, that individuals reproduce their like, that characteristics are inheritable—the likeness is a direct natural consequence of the genetic succession; "and it is logical to place the cause above the effect."

We are equally disposed to combat a proposition of De Candolle's about genera, elaborately argued in the "Géographie Botanique," and incidentally reaffirmed in his present article, viz., that genera are more natural than species, and more correctly distinguished by people in general, as is shown by vernacular names. But we have no space left in which to present some evidence to the contrary.

V.

SEQUOIA AND ITS HISTORY; THE RELATIONS OF NORTH AMERICAN TO NORTHEAST ASIAN AND TO TERTIARY VEGETATION.

(A PRESIDENTIAL ADDRESS TO THE AMERICAN ASSOCIATION FOR THE ADVANCEMENT OF SCIENCE. AT DUBUQUE, *August*, 1872.)

THE session being now happily inaugurated, your presiding officer of the last year has only one duty to perform before he surrenders the chair to his successor. If allowed to borrow a simile from the language of my own profession, I might liken the President of this Association to a biennial plant. He flourishes for the year in which he comes into existence, and performs his appropriate functions as presiding officer. When the second year comes round, he is expected to blossom out in an address and disappear. Each president, as he retires, is naturally expected to contribute something from his own investigations or his own line of study, usually to discuss some particular scientific topic.

Now, although I have cultivated the field of North American botany, with some assiduity, for more than forty years, have reviewed our vegetable hosts, and assigned to no small number of them their names and their place in the ranks, yet, so far as our own wide country is concerned, I have been to a great extent a closet botanist. Until this summer I had not seen the Mississippi, nor set foot upon a prairie.

To gratify a natural interest, and to gain some title for addressing a body of practical naturalists and explorers, I have made a pilgrimage across the continent. I have sought and viewed in their native haunts many a plant and flower which for me had long bloomed unseen, or only in the *hortus siccus*. I have been able to see for myself what species and what forms constitute the main features of the vegetation of each successive region, and record—as the vegetation unerringly does—the permanent characteristics of its climate.

Passing on from the eastern district, marked by its equably distributed rainfall, and therefore naturally forest-clad, I have seen the trees diminish in number, give place to wide prairies, restrict their growth to the borders of streams, and then disappear from the boundless drier plains; have seen grassy plains change into a brown and sere desert—desert in the common sense, but hardly anywhere botanically so—have seen a fair growth of coniferous trees adorning the more favored slopes of a mountain-range high enough to compel summer showers; have traversed that broad and bare elevated region shut off on both sides by high mountains from the moisture supplied by either ocean, and longitudinally intersected by sierras which seemingly remain as naked as they were born; and have reached at length the westward slopes of that high mountain-barrier which, refreshed by the Pacific, bears the noble forests of the Sierra Nevada and the Coast Ranges, and among them trees which are the wonder of the world. As I stood in their shade, in the groves of Mariposa and Calaveras, and again under the canopy of the commoner redwood, raised on columns of such majestic height and ample girth, it occurred to me that I could not do better than to share with you, upon this occasion, some of the thoughts which possessed my mind. In their development they may, perhaps, lead us up to questions of considerable scientific interest.

I shall not detain you with any remarks—which would now be trite—upon the size or longevity of these far-famed Sequoia-trees, or of the sugar-pines, incense-cedar, and firs associated with them, of which even the prodigious bulk of the dominating Sequoia does not sensibly diminish the grandeur. Although no account and no photographic representation of either species of the far-famed Sequoia-trees gives any adequate impression of their singular majesty—still less of their beauty—yet my interest in them did not culminate merely or mainly in considerations of their size and age. Other trees, in other parts of the world, may claim to be older. Certain Australian gum-trees (Eucalypti) are said to be taller. Some, we are told, rise so high that they might even cast a flicker of shadow upon the summit of the Pyramid of Cheops. Yet the oldest of them doubtless grew from seed which was shed long after the names of the pyramid-builders had been forgotten. So far as we can judge from the actual counting of the layers of several trees, no Sequoia now alive sensibly antedates the Christian era.

Nor was I much impressed with an attraction of

man's adding. That the more remarkable of these trees should bear distinguishing appellations seems proper enough; but the tablets of personal names which are affixed to many of them in the most visited groves—as if the memory of more or less notable people of our day might be made enduring by the juxtaposition—do suggest some incongruity. When we consider that a hand's breadth at the circumference of any one of the venerable trunks so placarded has recorded in annual lines the lifetime of the individual thus associated with it, one may question whether the next hand's breadth may not measure the fame of some of the names thus ticketed for adventitious immortality. Whether it be the man or the tree that is honored in the connection, probably either would live as long, in fact and in memory, without it.

One notable thing about the Sequoia-trees is their *isolation*. Most of the trees associated with them are of peculiar species, and some of them are nearly as local. Yet every pine, fir, and cypress of California is in some sort familiar, because it has near relatives in other parts of the world. But the redwoods have none. The redwood—including in that name the two species of "big-trees"—belongs to the general Cypress family, but is *sui generis*. Thus isolated systematically, and extremely isolated geographically, and so wonderful in size and port, they more than other trees suggest questions.

Were they created thus local and lonely, denizens of California only; one in limited numbers in a few choice spots on the Sierra Nevada, the other along the Coast Range from the Bay of Monterey to the frontiers of Oregon? Are they veritable Melchizedeks, without pedigree or early relationship, and possibly fated to be without descent?

Or are they now coming upon the stage—or rather were they coming but for man's interference—to play a part in the future?

Or are they remnants, sole and scanty survivors of a race that has played a grander part in the past, but is now verging to extinction? Have they had a career, and can that career be ascertained or surmised, so that we may at least guess whence they came, and how, and when?

Time was, and not long ago, when such questions as these were regarded as useless and vain—when students of natural history, unmindful of what the name denotes, were content with a knowledge of things as they now are, but gave little heed as to how they came to be so. Now such questions are held to be legitimate, and perhaps not wholly unanswerable. It cannot now be said that these trees inhabit their present restricted areas simply because they are there placed in the climate and soil of all the world most congenial to them. These must indeed be congenial, or they would not survive. But when we see how the Australian Eucalyptus-trees thrive upon the Californian coast, and how these very redwoods flourish upon another continent; how the so-called wild-oat (Avena sterilis of the Old World) has taken full possession of California; how that cattle and horses, introduced by the Spaniard, have spread as widely and made themselves as much at home on the plains of La Plata as on those of Tartary; and that the cardoon-thistle-seeds, and others they brought with them, have multiplied there into numbers probably much exceeding those extant in their native lands; indeed, when we contemplate our own race, and our particular stock, taking such recent but dominating possession of this New World; when we consider how the indigenous flora of islands generally succumbs to the foreigners which come in the train of man; and that most weeds (i. e., the prepotent plants in open soil) of all temperate climates are not "to the manner born," but are self-invited intruders—we must needs abandon the notion of any primordial and absolute adaptation of plants and animals to their habitats, which may stand in lieu of explanation, and so preclude our inquiring any further. The harmony of Nature and its admirable perfection need not be regarded as inflexible and changeless. Nor need Nature be likened to a statue, or a cast in rigid bronze, but rather to an organism, with play and adaptability of parts, and life and even soul informing the whole. Under the former view Nature would be "the faultless monster which the world ne'er saw," but inscrutable as the Sphinx, whom it were vain, or worse, to question of the whence and whither. Under the other, the perfection of Nature, if relative, is multifarious and ever renewed; and much that is enigmatical now may find explanation in some record of the past.

That the two species of redwood we are contemplating originated as they are and where they are, and for the part they are now playing, is, to say the least, not a scientific supposition, nor in any sense a probable one. Nor is it more likely that they are destined to play a conspicuous part in the future, or that they would have done so, even if the Indian's fires and the white man's axe had spared them. The redwood of the coast (*Sequoia sempervirens*) had the stronger hold upon existence, forming as it did large forests throughout a narrow belt about three hundred miles in length, and being so tenacious of life that every large stump sprouts into a copse. But it does not pass the bay of Monterey, nor cross the line of Oregon, although so grandly developed not far below it. The more remarkable *Sequoia gigantea* of the Sierra exists in numbers so limited that the separate groves may be reckoned upon the fingers, and the trees of most of them have been counted, except near their southern limit, where they are said to be more copious. A species limited in individuals holds its existence by a precarious tenure; and this has a foothold only in a few sheltered spots, of a happy mean in temperature, and locally favored with moisture in summer. Even there, for some reason or other, the pines with which they are associated (Pinus Lambertiana and P. ponderosa), the firs (Abies grandis and A. amabilis), and even the incense-cedar (Libocedrus decurrens), possess a great advantage, and, though they strive in vain to emulate their size, wholly overpower the Sequoias in numbers. "To him that hath shall be given." The force of numbers eventually wins. At least in the commonly-visited groves Sequoia gigantea is invested in its last stronghold, can neither advance into more exposed positions above, nor fall back into drier and barer ground below, nor hold its own in the long-run where it is, under present conditions; and a little further drying of the climate, which must once have been much moister than now, would precipitate its doom. Whatever the individual longevity, certain if not speedy is the decline of a race in which a high death-rate afflicts the young. Seedlings of the big trees occur not rarely, indeed, but in meagre proportion to those of associated trees; and small indeed is the chance that any of these will attain to "the days of the years of their fathers." "Few and evil" are the days of all the forest likely to be, while man, both barbarian and civilized, torments them with fires, fatal at once to seedlings, and at length to the aged also. The forests of California, proud as the State may be of them, are already too scanty and insufficient for her uses. Two lines, such as may be drawn with one sweep of a brush over the map, would cover them all. The coast redwood—the most important tree in California, although a million times more numerous than its relative of the Sierra—is too good to live long. Such is its value for lumber and its accessibility, that, judging the future by the past, it is not likely, in its primeval growth, to outlast its rarer fellow-species.

Happily man preserves and disseminates as well as destroys. The species will doubtless be preserved to science, and for ornamental and other uses, in its own and other lands; and the more remarkable individuals of the present day are likely to be sedulously cared for, all the more so as they become scarce.

Our third question remains to be answered: Have these famous Sequoias played in former times and upon a larger stage a more imposing part, of which the present is but the epilogue? We cannot gaze high up the huge and venerable trunks, which one crosses the continent to behold, without wishing that these patriarchs of the grove were able, like the long-lived antediluvians of Scripture, to hand down to us, through a few generations, the traditions of centuries, and so tell us somewhat of the history of their race. Fifteen hundred annual layers have been counted, or satisfactorily made out, upon one or two fallen trunks. It is probable that close to the heart of some of the living trees may be found the circle that records the year of our Saviour's nativity. A few generations of such trees might carry the history a long way back. But the ground they stand upon, and the marks of very recent geological change and vicissitude in the region around, testify that not very many such generations can have flourished just there, at least in an unbroken series. When their site was covered by glaciers, these Sequoias must have occupied other stations, if, as there is reason to believe, they then existed in the land.

I have said that the redwoods have no near relatives in the country of their abode, and none of their genus anywhere else. Perhaps something may be learned of their genealogy by inquiring of such relatives as they have. There are only two of any particular nearness of kin; and they are far away. One is the bald cypress, our Southern cypress, *Taxodium*, inhabiting the swamps of the Atlantic coast from Maryland to Texas, thence extending—with, probably, a specific difference—into Mexico. It is well known as one of the largest trees of our Atlantic forest-district, and, although it never—except perhaps in Mexico, and in rare instances—attains the portliness of its Western relatives, yet it may equal them in longevity. The other relative is *Glyptostrobus*, a sort of modified Taxodium, being about as much like our bald cypress as one species of redwood is like the other.

Now, species of the same type, especially when few, and the type peculiar, are, in a general way, associated geographically, i. e., inhabit the same country, or (in a large sense) the same region. Where it is not so, where near relatives are separated, there is usually something to be explained. Here is an instance. These four trees, sole representatives of their tribe, dwell almost in three separate quarters of the world: the two redwoods in California, the bald cypress in Atlantic North America, its near relative, Glyptostrobus, in China.

It was not always so. In the Tertiary period, the geological botanists assure us, our own very Taxodium or bald cypress, and a Glyptostrobus, exceedingly like the present Chinese tree, and more than one Sequoia, coexisted in a fourth quarter of the globe, viz., in Europe! This brings up the question, Is it possible to bridge over these four wide intervals of space and the much vaster interval of time, so as to bring these extraordinarily separated relatives into connection? The evidence which may be brought to bear upon this question is various and widely scattered. I bespeak your patience while I endeavor to bring together, in an abstract, the most important points of it.

Some interesting facts may come out by comparing generally the botany of the three remote regions, each of which is the sole home of one of these genera, i. e., Sequoia in California, Taxodium in the Atlantic United

States,[1] and Glyptostrobus in China, which compose the whole of the peculiar tribe under consideration.

Note then, first, that there is another set of three or four peculiar trees, in this case of the yew family, which has just the same peculiar distribution, and which therefore may have the same explanation, whatever that explanation be. The genus *Torreya,* which commemorates our botanical Nestor and a former president of this Association, Dr. Torrey, was founded upon a tree rather lately discovered (that is, about thirty-five years ago) in Northern Florida. It is a noble, yew-like tree, and very local, being, so far as known, nearly confined to a few miles along the shores of a single river. It seems as if it had somehow been crowded down out of the Alleghanies into its present limited southern quarters; for in cultivation it evinces a northern hardiness. Now, another species of Torreya is a characteristic tree of Japan; and one very like it, if not the same, inhabits the mountains of Northern China—belongs, therefore, to the Eastern Asiatic temperate region, of which Northern China is a part, and Japan, as we shall see, the portion most interesting to us. There is only one more species of Torreya, and that is a companion of the redwoods in California. It is the tree locally known under the name of the California nutmeg. Here are three or four near brethren, species of the same genus, known nowhere else than in these three habitats.

[1] The phrase "Atlantic United States" is here used throughout in contradistinction to Pacific United States: to the former of course belong, botanically and geographically, the valley of the Mississippi and its tributaries up to the eastern border of the great woodless plains, which constitute an intermediate region.

Moreover, the Torreya of Florida is associated with a yew; and the trees of this grove are the only yew-trees of Eastern North America; for the yew of our Northern woods is a decumbent shrub. A yew-tree, perhaps the same, is found with Taxodium in the temperate parts of Mexico. The only other yews in America grow with the redwoods and the other Torreya in California, and extend northward into Oregon. Yews are also associated with Torreya in Japan; and they extend westward through Mantchooria and the Himalayas to Western Europe, and even to the Azores Islands, where occurs the common yew of the Old World.

So we have three groups of coniferous trees which agree in this peculiar geographical distribution, with, however, a notable extension of range in the case of the yew: 1. The redwoods, and their relatives, Taxodium and Glyptostrobus, which differ so as to constitute a genus for each of the three regions; 2. The Torreyas, more nearly akin, merely a different species in each region; 3. The yews, still more closely related while more widely disseminated, of which it is yet uncertain whether they constitute seven, five, three, or only one species. Opinions differ, and can hardly be brought to any decisive test. However it be determined, it may still be said that the extreme differences among the yews do not surpass those of the recognized variations of the European yew, the cultivated races included.

It appears to me that these several instances all raise the very same question, only with different degrees of emphasis, and, if to be explained at all, will have the same kind of explanation.

Continuing the comparison between the three regions with which we are concerned, we note that each has its own species of pines, firs, larches, etc., and of a few deciduous-leaved trees, such as oaks and maples; all of which have no peculiar significance for the present purpose, because they are of genera which are common all round the northern hemisphere. Leaving these out of view, the noticeable point is that the vegetation of California is most strikingly unlike that of the Atlantic United States. They possess some plants, and some peculiarly American plants, in common—enough to show, as I imagine, that the difficulty was not in the getting from the one district to the other, or into both from a common source, but in abiding there. The primordially unbroken forest of Atlantic North America, nourished by rainfall distributed throughout the year, is widely separated from the western region of sparse and discontinuous tree-belts of the same latitude on the western side of the continent (where summer rain is wanting, or nearly so), by immense treeless plains and plateaux of more or less aridity, traversed by longitudinal mountain-ranges of a similar character. Their nearest approach is at the north, in the latitude of Lake Superior, where, on a more rainy line, trees of the Atlantic forest and that of Oregon may be said to intermix. The change of species and of the aspect of vegetation in crossing, say on the forty-seventh parallel, is slight in comparison with that on the thirty-seventh or near it. Confining our attention to the lower latitude, and under the exceptions already specially noted, we may say that almost every characteristic form in the vegetation of the Atlantic States is wanting in California, and the characteristic plants and trees of California are wanting here.

California has no magnolia nor tulip trees, nor star-anise tree; no so-called papaw (Asimina); no barberry of the common single-leaved sort; no Podophyllum or other of the peculiar associated genera; no nelumbo nor white water-lily; no prickly ash nor sumach; no loblolly-bay nor Stuartia; no basswood nor linden-trees; neither locust, honey-locust, coffee-trees (Gymnocladus) nor yellow-wood (Cladrastis); nothing answering to Hydrangea or witch-hazel, to gum-trees (Nyssa and Liquidambar), Viburnum or Diervilla; it has few asters and golden-rods; no lobelias; no huckleberries and hardly any blueberries; no Epigæa, charm of our earliest Eastern spring, tempering an icy April wind with a delicious wild fragrance; no Kalmia nor Clethra, nor holly, nor persimmon; no catalpa-tree, nor trumpet-creeper (Tecoma); nothing answering to sassafras, nor to benzoin-tree, nor to hickory; neither mulberry nor elm; no beech, true chestnut, hornbeam, nor ironwood, nor a proper birch-tree; and the enumeration might be continued very much further by naming herbaceous plants and others familiar only to botanists.

In their place California is filled with plants of other types—trees, shrubs, and herbs, of which I will only remark that they are, with one or two exceptions, as different from the plants of the Eastern Asiatic region with which we are concerned (Japan, China, and

Mantchooria), as they are from those of Atlantic North America. Their near relatives, when they have any in other lands, are mostly southward, on the Mexican plateau, or many as far south as Chili. The same may be said of the plants of the intervening great Plains, except that northward in the subsaline vegetation there are some close alliances with the flora of the steppes of Siberia. And along the crests of high mountain-ranges the Arctic-Alpine flora has sent southward more or less numerous representatives through the whole length of the country.

If we now compare, as to their flora generally, the Atlantic United States with Japan, Mantchooria, and Northern China—i. e., Eastern North America with Eastern North Asia, half the earth's circumference apart—we find an astonishing similarity. The larger part of the genera of our own region, which I have enumerated as wanting in California, are present in Japan or Mantchooria, along with many other peculiar plants, divided between the two. There are plants enough of the one region which have no representatives in the other. There are types which appear to have reached the Atlantic States from the south; and there is a larger infusion of subtropical Asiatic types into temperate China and Japan; among these there is no relationship between the two countries to speak of. There are also, as I have already said, no small number of genera and some species which, being common all round or partly round the northern temperate zone, have no special significance because of their occurrence in these two antipodal floras, although they have testimony to bear upon the general question of geographical distribution. The point to be remarked is, that many, or even most, of the genera and species which are peculiar to North America as compared with Europe, and largely peculiar to Atlantic North America as compared with the Californian region, are also represented in Japan and Mantchooria, either by identical or by closely-similar forms! The same rule holds on a more northward line, although not so strikingly. If we compare the plants, say of New England and Pennsylvania (latitude 45°–47°), with those of Oregon, and then with those of Northeastern Asia, we shall find many of our own curiously repeated in the latter, while only a small number of them can be traced along the route even so far as the western slope of the Rocky Mountains. And these repetitions of East American types in Japan and neighboring districts are in all degrees of likeness. Sometimes the one is undistinguishable from the other; sometimes there is a difference of aspect, but hardly of tangible character; sometimes the two would be termed marked varieties if they grew naturally in the same forest or in the same region; sometimes they are what the botanist calls representative species, the one answering closely to the other, but with some differences regarded as specific; sometimes the two are merely of the same genus, or not quite that, but of a single or very few species in each country; in which case the point which interests us is, that this peculiar limited type should occur in two antipodal places, and nowhere else.

It would be tedious, and, except to botanists, abstruse, to enumerate instances; yet the whole strength of the case depends upon the number of such instances. I propose therefore, if the Association does me the honor to print this discourse, to append in a note a list of the more remarkable ones.[1] But I would here mention certain cases as specimens.

Our Rhus Toxicodendron, or poison-ivy, is very exactly repeated in Japan, but is found in no other part of the world, although a species much like it abounds in California. Our other poisonous Rhus (R. venenata), commonly called poison-dogwood, is in no way represented in Western America, but has so close an analogue in Japan that the two were taken for the same by Thunberg and Linnæus, who called them both R. vernix.

Our northern fox-grape, Vitis Labrusca, is wholly confined to the Atlantic States, except that it reappears in Japan and that region.

The original Wistaria is a woody leguminous climber with showy blossoms, native to the middle Atlantic States; the other species, which we so much prize in cultivation, W. Sinensis, is from China, as its name denotes, or perhaps only from Japan, where it is certainly indigenous.

Our yellow-wood (Cladrastis) inhabits a very limited district on the western slope of the Alleghanies. Its only and very near relative, Maackia, is confined to Mantchooria.

The Hydrangeas have some species in our Alleghany region: all the rest belong to the Chino-Japanese region and its continuation westward. The same may be said of Philadelphus, except that there are one

[1] The tabulated list referred to was printed as an appendix to the official edition of this discourse, but is here omitted.

or two mostly very similar species in California and Oregon.

Our May-flower (Epigæa) and our creeping snowberry, otherwise peculiar to Atlantic North America, recur in Japan.

Our blue cohosh (Caulophyllum) is confined to the woods of the Atlantic States, but has lately been discovered in Japan. A peculiar relative of it, Diphylleia, confined to the higher Alleghanies, is also repeated in Japan, with a slight difference, so that it may barely be distinguished as another species. Another relative is our twin-leaf (Jeffersonia) of the Alleghany region alone: a second species has lately turned up in Mantchooria. A relative of this is Podophyllum, our mandrake, a common inhabitant of the Atlantic United States, but found nowhere else. There is one other species of it, and that is in the Himalayas. Here are four most peculiar genera of one family, each of a single species in the Atlantic United States, which are duplicated on the other side of the world, either in identical or almost identical species, or in an analogous species, while nothing else of the kind is known in any other part of the world.

I ought not to omit ginseng, the root so prized by the Chinese, which they obtained from their northern provinces and Mantchooria, and which is now known to inhabit Corea and Northern Japan. The Jesuit Fathers identified the plant in Canada and the Atlantic States, brought over the Chinese name by which we know it, and established the trade in it, which was for many years most profitable. The exportation of ginseng to China probably has not yet entirely ceased.

Whether the Asiatic and the Atlantic American gin-sengs are to be regarded as of the same species or not is somewhat uncertain, but they are hardly, if at all, distinguishable.

There is a shrub, Elliottia, which is so rare and local that it is known only at two stations on the Savannah River in Georgia. It is of peculiar structure, and was without near relative until one was lately discovered in Japan (Tripetaleia), so like it as hardly to be distinguishable except by having the parts of the blossom in threes instead of fours—a difference not uncommon in the same genus, or even in the same species.

Suppose Elliottia had happened to be collected only once, a good while ago, and all knowledge of the limited and obscure locality were lost; and meanwhile the Japanese form came to be known. Such a case would be parallel with an actual one. A specimen of a peculiar plant (Shortia galacifolia) was detected in the herbarium of the elder Michaux, who collected it (as his autograph ticket shows) somewhere in the high Alleghany Mountains, more than eighty years ago. No one has seen the living plant since or knows where to find it, if haply it still flourishes in some secluded spot. At length it is found in Japan; and I had the satisfaction of making the identification.[1] A relative is also known in Japan; and a less near one has just been detected in Thibet.

Whether the Japanese and the Alleghanian plants are exactly the same or not, it needs complete specimens of the two to settle. So far as we know, they

[1] *American Journal of Science*, 1867, p. 402; "Proceedings of American Academy," vol. viii., p. 244.

are just alike; and, even if some difference were discerned between them, it would not appreciably alter the question as to how such a result came to pass. Each and every one of the analogous cases I have been detailing—and very many more could be mentioned—raises the same question, and would be satisfied with the same answer.

These singular relations attracted my curiosity early in the course of my botanical studies, when comparatively few of them were known, and my serious attention in later years, when I had numerous and new Japanese plants to study in the collections made, by Messrs. Williams and Morrow, during Commodore Perry's visit in 1853, and especially, by Mr. Charles Wright, of Commodore Rodgers's expedition in 1855. I then discussed this subject somewhat fully, and tabulated the facts within my reach.[1]

This was before Heer had developed the rich fossil botany of the arctic zone, before the immense antiquity of existing species of plants was recognized, and before the publication of Darwin's now famous volume on the "Origin of Species" had introduced and familiarized the scientific world with those now current ideas respecting the history and vicissitudes of species with which I attempted to deal in a moderate and feeble way.

My speculation was based upon the former glaciation of the northern temperate zone, and the inference of a warmer period preceding and perhaps following. I considered that our own present vegetation, or its proximate ancestry, must have occupied the arctic and

[1] "Memoirs of American Academy," vol. vi., pp. 377–458 (1859).

subarctic regions in pliocene times, and that it had been gradually pushed southward as the temperature lowered and the glaciation advanced, even beyond its present habitation; that plants of the same stock and kindred, probably ranging round the arctic zone as the present arctic species do, made their forced migration southward upon widely different longitudes, and receded more or less as the climate grew warmer; that the general difference of climate which marks the eastern and the western sides of the continents—the one extreme, the other mean—was doubtless even then established, so that the same species and the same sorts of species would be likely to secure and retain foothold in the similar climates of Japan and the Atlantic United States, but not in intermediate regions of different distribution of heat and moisture; so that different species of the same genus, as in Torreya, or different genera of the same group, as redwood, Taxodium, and Glyptostrobus, or different associations of forest-trees, might establish themselves each in the region best suited to the particular requirements, while they would fail to do so in any other. These views implied that the sources of our actual vegetation and the explanation of these peculiarities were to be sought in, and presupposed, an ancestry in pliocene or earlier times, occupying the higher northern regions. And it was thought that the occurrence of peculiar North American genera in Europe in the Tertiary period (such as Taxodium, Carya, Liquidambar, sassafras, Negundo, etc.) might be best explained on the assumption of early interchange and diffusion through North Asia, rather than by that of the fabled Atlantis.

The hypothesis supposed a gradual modification of species in different directions under altering conditions, at least to the extent of producing varieties, sub-species, and representative species, as they may be variously regarded; likewise the single and local origination of each type, which is now almost universally taken for granted.

The remarkable facts in regard to the Eastern American and Asiatic floras which these speculations were to explain have since increased in number, especially through the admirable collections of Dr. Maximowicz in Japan and adjacent countries, and the critical comparisons he has made and is still engaged upon.

I am bound to state that, in a recent general work[1] by a distinguished European botanist, Prof. Grisebach, of Göttingen, these facts have been emptied of all special significance, and the relations between the Japanese and the Atlantic United States flora declared to be no more intimate than might be expected from the situation, climate, and present opportunity of interchange. This extraordinary conclusion is reached by regarding as distinct species all the plants common to both countries between which any differences have been discerned, although such differences would probably count for little if the two inhabited the same country, thus transferring many of my list of identical to that of representative species; and then by simply eliminating from consideration the whole array of representative species, i. e., all cases in which the Japanese and the American plant are not exactly alike.

[1] "Die Vegetation der Erde nach ihrer klimatischen Anordnung," 1871.

As if, by pronouncing the cabalistic word *species*, the question were settled, or rather the greater part of it remanded out of the domain of science; as if, while complete identity of forms implied community of origin, anything short of it carried no presumption of the kind; so leaving all these singular duplicates to be wondered at, indeed, but wholly beyond the reach of inquiry.

Now, the only known cause of such likeness is inheritance; and as all transmission of likeness is with some difference in individuals, and as changed conditions have resulted, as is well known, in very considerable differences, it seems to me that, if the high antiquity of our actual vegetation could be rendered probable, not to say certain, and the former habitation of any of our species or of very near relatives of them in high northern regions could be ascertained, my whole case would be made out. The needful facts, of which I was ignorant when my essay was published, have now been for some years made known—thanks, mainly, to the researches of Heer upon ample collections of arctic fossil plants. These are confirmed and extended by new investigations, by Heer and Lesquereux, the results of which have been indicated to me by the latter.[1]

The Taxodium, which everywhere abounds in the

[1] Reference should also be made to the extensive researches of Newberry upon the tertiary and cretaceous floras of the Western United States. *See* especially Prof. Newberry's paper in the *Boston Journal of Natural History*, vol. vii., No. 4, describing fossil plants of Vancouver's Island, etc.; his "Notes on the Later Extinct Floras of North America," etc., in "Annals of the Lyceum of Natural History," vol. ix., April, 1868; "Report on the Cretaceous and Tertiary Plants

miocene formations in Europe, has been specifically identified, first by Gœppert, then by Heer, with our common cypress of the Southern States. It has been found fossil in Spitzbergen, Greenland, and Alaska—in the latter country along with the remains of another form, distinguishable, but very like the common species; and this has been identified by Lesquereux in the miocene of the Rocky Mountains. So there is one species of tree which has come down essentially unchanged from the Tertiary period, which for a long while inhabited both Europe and North America, and also, at some part of the period, the region which geographically connects the two (once doubtless much more closely than now), but which has survived only in the Atlantic United States and Mexico.

The same Sequoia which abounds in the same mio cene formations in Northern Europe has been abundantly found in those of Iceland, Spitzbergen, Greenland, Mackenzie River, and Alaska. It is named *S. Langsdorfii*, but is pronounced to be very much like *S. sempervirens*, our living redwood of the Californian coast, and to be the ancient representative of it. Fossil specimens of a similar, if not the same, species have recently been detected in the Rocky Mountains by Hayden, and determined by our eminent palæontological botanist, Lesquereux; and he assures me that he has

collected in Raynolds and Hayden's Yellowstone and Missouri Exploring Expedition, 1859–1860," published in 1869; and an interesting article entitled "The Ancient Lakes of Western America, their Deposits and Drainage," published in *The American Naturalist*, January, 1871.

The only document I was able to consult was Lesquereux's "Report on the Fossil Plants," in Hayden's report of 1872.

the common redwood itself from Oregon in a deposit of tertiary age. Another Sequoia (*S. Sternbergii*), discovered in miocene deposits in Greenland, is pronounced to be the representative of *S. gigantea*, the big tree of the Californian Sierra. If the Taxodium of the tertiary time in Europe and throughout the arctic regions is the ancestor of our present bald cypress—which is assumed in regarding them as specifically identical—then I think we may, with our present light, fairly assume that the two redwoods of California are the direct or collateral descendants of the two ancient species which so closely resemble them.

The forests of the arctic zone in tertiary times contained at least three other species of Sequoia, as determined by their remains, one of which, from Spitzbergen, also much resembles the common redwood of California. Another, "which appears to have been the commonest coniferous tree on Disco," was common in England and some other parts of Europe. So the Sequoias, now remarkable for their restricted station and numbers, as well as for their extraordinary size, are of an ancient stock; their ancestors and kindred formed a large part of the forests which flourished throughout the polar regions, now desolate and ice-clad, and which extended into low latitudes in Europe. On this continent one species, at least, had reached to the vicinity of its present habitat before the glaciation of the region. Among the fossil specimens already found in California, but which our trustworthy palæontological botanist has not yet had time to examine, we may expect to find evidence of the early arrival of these two redwoods upon the ground which they now, after much vicissitude, scantily occupy.

Differences of climate, or circumstances of migration, or both, must have determined the survival of Sequoia upon the Pacific, and of Taxodium upon the Atlantic coast. And still the redwoods will not stand in the east, nor could our Taxodium find a congenial station in California. Both have probably had their opportunity in the olden time, and failed.

As to the remaining near relative of Sequoia, the Chinese Glyptostrobus, a species of it, and its veritable representative, was contemporaneous with Sequoia and Taxodium, not only in temperate Europe, but throughout the arctic regions from Greenland to Alaska. According to Newberry, it was abundantly represented in the miocene flora of the temperate zone of our own continent, from Nebraska to the Pacific.

Very similar would seem to have been the fate of a more familiar gymnospermous tree, the Gingko or Salisburia. It is now indigenous to Japan only. Its ancestor, as we may fairly call it—since, according to Heer, "it corresponds so entirely with the living species that it can scarcely be separated from it"—once inhabited Northern Europe and the whole arctic region round to Alaska, and had even a representative farther south, in our Rocky Mountain district. For some reason, this and Glyptostrobus survive only on the shores of Eastern Asia.

Libocedrus, on the other hand, appears to have cast in its lot with the Sequoias. Two species, according to Heer, were with them in Spitzbergen. *L. decurrens*, the incense cedar, is one of the noblest associates of the present redwoods. But all the rest are in the southern hemisphere, two at the southern extremity of the Andes, two in the South-Sea Islands. It is only by bold and far-reaching suppositions that they can be geographically associated.

The genealogy of the Torreyas is still wholly obscure; yet it is not unlikely that the yew-like trees, named Taxites, which flourished with the Sequoias in the tertiary arctic forests, are the remote ancestors of the three species of Torreya, now severally in Florida, in California, and in Japan.

As to the pines and firs, these were more numerously associated with the ancient Sequoias of the polar forests than with their present representatives, but in different species, apparently more like those of Eastern than of Western North America. They must have encircled the polar zone then, as they encircle the present temperate zone now.

I must refrain from all enumeration of the angiospermous or ordinary deciduous trees and shrubs, which are now known, by their fossil remains, to have flourished throughout the polar regions when Greenland better deserved its name and enjoyed the present climate of New England and New Jersey. Then Greenland and the rest of the north abounded with oaks, representing the several groups of species which now inhabit both our Eastern and Western forest districts; several poplars, one very like our balsam poplar or balm-of-Gilead tree; more beeches than there are now, a hornbeam, and a hop-hornbeam, some birches, a persimmon, and a planer-tree, near representatives of those of the Old World, at least of

Asia, as well as of Atlantic North America, but all wanting in California; one Juglans like the walnut of the Old World, and another like our black walnut; two or three grapevines, one near our Southern fox grape or muscadine, another near our Northern frost-grape; a Tilia, very like our basswood of the Atlantic States only; a Liquidambar; a magnolia, which recalls our M. grandiflora; a Liriodendron, sole representative of our tulip-tree; and a sassafras, very like the living tree.

Most of these, it will be noticed, have their nearest or their only living representatives in the Atlantic States, and when elsewhere, mainly in Eastern Asia. Several of them, or of species like them, have been detected in our tertiary deposits, west of the Mississippi, by Newberry and Lesquereux. Herbaceous plants, as it happens, are rarely preserved in a fossil state, else they would probably supply additional testimony to the antiquity of our existing vegetation, its wide diffusion over the northern and now frigid zone, and its enforced migration under changes of climate.[1]

Concluding, then, as we must, that our existing vegetation is a continuation of that of the tertiary

[1] There is, at least, one instance so opportune to the present argument that it should not pass unnoticed, although I had overlooked the record until now. *Onoclea sensibilis* is a fern peculiar to the Atlantic United States (where it is common and wide-spread) and to Japan. Prof. Newberry identified it several years ago in a collection, obtained by Dr. Hayden, of miocene fossil plants of Dakota Territory, which is far beyond it present habitat. He moreover regards it as probably identical with a fossil specimen "described by the late Prof. E. Forbes, under the name of *Filicites Hebridicus*, and obtained by the Duke of Argyll from the island of Mull."

period, may we suppose that it absolutely originated then? Evidently not. The preceding Cretaceous period has furnished to Carruthers in Europe a fossil fruit like that of the *Sequoia gigantea* of the famous groves, associated with pines of the same character as those that accompany the present tree; has furnished to Heer, from Greenland, two more Sequoias, one of them identical with a tertiary species, and one nearly allied to *Sequoia Langsdorfii*, which in turn is a probable ancestor of the common Californian redwood; has furnished to Newberry and Lesquereux in North America the remains of another ancient Sequoia, a Glyptostrobus, a Liquidambar which well represents our sweet-gum-tree, oaks analogous to living ones, leaves of a plane-tree, which are also in the Tertiary, and are scarcely distinguishable from our own *Platanus occidentalis*, of a magnolia and a tulip-tree, and "of a sassafras undistinguishable from our living species." I need not continue the enumeration. Suffice it to say that the facts justifiy the conclusion which Lesquereux —a scrupulous investigator—has already announced: that "the essential types of our actual flora are marked in the Cretaceous period, and have come to us after passing, without notable changes, through the Tertiary formations of our continent."

According to these views, as regards plants at least, the adaptation to successive times and changed conditions has been maintained, not by absolute renewals, but by gradual modifications. I, for one, cannot doubt that the present existing species are the lineal successors of those that garnished the earth in the old time before them, and that they were as well adapted to their surroundings then, as those which flourish and bloom around us are to their conditions now. Order and exquisite adaptation did not wait for man's coming, nor were they ever stereotyped. Organic Nature—by which I mean the system and totality of living things, and their adaptation to each other and to the world—with all its apparent and indeed real stability, should be likened, not to the ocean, which varies only by tidal oscillations from a fixed level to which it is always returning, but rather to a river, so vast that we can neither discern its shores nor reach its sources, whose onward flow is not less actual because too slow to be observed by the *ephemeræ* which hover over its surface, or are borne upon its bosom.

Such ideas as these, though still repugnant to some, and not long since to many, have so possessed the minds of the naturalists of the present day that hardly a discourse can be pronounced or an investigation prosecuted without reference to them. I suppose that the views here taken are little, if at all, in advance of the average scientific mind of the day. I cannot regard them as less noble than those which they are succeeding. An able philosophical writer, Miss Frances Power Cobbe, has recently and truthfully said:[1]

> "It is a singular fact that, when we can find out how anything is done, our first conclusion seems to be that God did not do it. No matter how wonderful, how beautiful, how intimately complex and delicate has been the machinery which has worked, perhaps for centuries, perhaps for millions of ages, to bring about some beneficent result, if we can but catch a glimpse of the wheels its divine character disappears."

[1] "Darwinism in Morals," in *Theological Review*, April, 1871.

I agree with the writer that this first conclusion is premature and unworthy — I will add, deplorable. Through what faults or infirmities of dogmatism on the one hand, and skepticism on the other, it came to be so thought, we need not here consider. Let us hope, and I confidently expect, that it is not to last; that the religious faith which survived without a shock the notion of the fixity of the earth itself may equally outlast the notion of the fixity of the species which inhabit it; that, in the future even more than in the past, faith in an *order*, which is the basis of science, will not—as it cannot reasonably—be dissevered from faith in an *Ordainer*, which is the basis of religion.

VI.

THE ATTITUDE OF WORKING NATURALISTS TOWARD DARWINISM.[1]

(The Nation, *October* 16, 1873.)

That homely adage, "What is one man's meat is another man's poison," comes to mind when we consider with what different eyes different naturalists look upon the hypothesis of the derivative origin of actual specific forms, since Mr. Darwin gave it vogue and

[1] "Histoire des Sciences et des Sevants depuis deux Siècles, suivie d'autres études sur des sujets scientifiques, en particulier sur la Sélection dans l'Espèce Humaine, par Alphonse De Candolle." Genève: H. Georg. 1873.

"Addresses of George Bentham, President, read at the anniversary meetings of the Linnean Society, 1862–1873."

"Notes on the Classification, History, and Geographical Distribution of Compositæ. By George Bentham." Separate issue from the Journal of the Linnean Society. Vol. XIII. London. 1873.

"On Palæontological Evidence of Gradual Modification of Animal Forms, read at the Royal Institution of Great Britain, April 25, 1873. by Prof. W. H. Flower." (*Journal of the Royal Institution*, pp. 11.)

"The Distribution and Migration of Birds. Memoir presented to the National Academy of Sciences, January, 1865, abstracted in the *American Journal of Science and the Arts*. 1866, etc. By Spencer F. Baird."

"The Story of the Earth and Man. By J. W. Dawson, LL. D., F. R. S., F. G. S., Principal and Vice-Chancellor of McGill University, Montreal. London: Hodder & Stoughton; New York: Harper & Brothers. 1873. Pp. 403, 12mo.

vigor and a *raison d'être* for the present day. This latter he did, not only by bringing forward a *vera causa* in *the survival of the fittest* under changing circumstances—about which the question among naturalists mainly is how much it will explain, some allowing it a restricted, others an unlimited operation—but also by showing that the theory may be made to do work, may shape and direct investigations, the results of which must in time tell us whether the theory is likely to hold good or not. If the hypothesis of natural selection and the things thereto appertaining had not been capable of being put to useful work, although, like the "Vestiges of the Natural History of Creation," it might have made no little noise in the world, it would hardly have engaged the attention of working naturalists as it has done. We have no idea even of opening the question as to what work the Darwinian theory has incited, and in what way the work done has reacted upon the theory; and least of all do we like to meddle with the polemical literature of the subject, already so voluminous that the German bibliographers and booksellers make a separate class of it. But two or three treatises before us, of a minor or incidental sort, suggest a remark or two upon the attitude of mind toward evolutionary theories taken by some of the working naturalists.

Mr. Darwin's own expectation, that his new presentation of the subject would have little or no effect upon those who had already reached middle-age, has —out of Paris—not been fulfilled. There are, indeed, one or two who have thought it their duty to denounce the theory as morally dangerous, as well as scientifically baseless; a recent instance of the sort we may have to consider further on. Others, like the youth at the river's bank, have been waiting in confident expectation of seeing the current run itself dry. On the other hand, a notable proportion of the more active-minded naturalists had already come to doubt the received doctrine of the entire fixity of species, and still more that of their independent and supernatural origination. While their systematic work all proceeded implicitly upon the hypothesis of the independence and entire permanence of species, they were perceiving more or less clearly that the whole question was inevitably to be mooted again, and so were prepared to give the alternative hypothesis a dispassionate consideration. The veteran Lyell set an early example, and, on a reconsideration of the whole question, wrote anew his famous chapter and reversed his former and weighty opinion. Owen, still earlier, signified his adhesion to the doctrine of derivation in some form, but apparently upon general, speculative grounds; for he repudiated natural selection, and offered no other natural solution of the mystery of the orderly incoming of cognate forms. As examples of the effect of Darwin's "Origin of Species" upon the minds of naturalists who are no longer young, and whose prepossessions, even more than Lyell's, were likely to bias them against the new doctrine, two from the botanical side are brought to our notice through recent miscellaneous writings which are now before us.[1]

[1] Since this article was in type, noteworthy examples of appreciative scientific judgment of the derivative hypothesis have come to hand: 1. In the opening address to the Geological Section of the British Associa-

Before the publication of Darwin's first volume, M. Alphonse de Candolle had summed up the result of his studies in this regard, in the final chapter of his classical "Géographie Botanique Raisonnée," in the conclusion, that existing vegetation must be regarded as the continuation, through many geological and geographical changes, of the anterior vegetations of the world; and that, consequently, the present distribution of species is explicable only in the light of their geological history. He surmised that, notwithstanding the general stability of forms, certain species or quasi-species might have originated through diversification under geographical isolation. But, on the other hand, he was still disposed to admit that even the same species might have originated independently in two or more different regions of the world; and he declined, as unpractical and unavailing, all attempts to apply hypotheses to the elucidation of the origin of species. Soon after Darwin's book appeared, De Candolle had occasion to study systematically a large and wide-spread genus—that of the oak. Investigating it under the new light of natural selection, he came to the conclusion that the existing oaks are all descendants of earlier forms, and that no clear line can be drawn between the diversification which has

tion, at its recent meeting, by its president, the veteran Phillips, perhaps the oldest surviving geologist after Lyell; and, 2. That of Prof. Allman, President of the Biological Section. The first touches the subject briefly, but in the way of favorable suggestion; the second is a full and discriminating exposition of the reasons which seem to assure at least the provisional acceptance of the hypothesis, as a guide in all biological studies, "a key to the order and hidden forces of the world of life."

resulted in species and that which is exhibited in races and minor varieties.

And now, in the introductory chapter of the volume of essays before us, he informs us that the idea which pervades them all, and in some sort connects very diverse topics, is that of considering this principle of selection. Of the principle itself, he remarks that it is neither a theory nor an hypothesis, but the expression of a necessary fact; that to deny it is very much like denying that round stones will roll downhill faster and farther than flat ones; and that the question of the present day in natural history is not whether there be natural selection, or even whether forms are derived from other forms, but to comprehend how, in what proportions, and by what means hereditary deviations take place, and in what ways an inevitable selection takes effect upon these. In two of these essays natural selection is directly discussed in its application to the human race; the larger one dealing ably with the whole subject, and with results at first view seemingly in a great degree negative, but yet showing that the supposed "failure of natural selection in the case of man" was an unwarrantable conclusion from too limited a view of a very complicated question. The article abounds in acute and fertile suggestions, and its closing chapter, "on the probable future of the human species" under the laws of selection, is highly interesting and noteworthy. The other and shorter essay discusses a special point, and brings out a corollary of the law of heredity which may not have been thought of before, but which is perfectly clear as soon as it is stated. It explains at once why contagious or epidemic diseases are most fatal at their first appearance, and less so afterward: not by the dying out of a virus—for, when the disease reaches a new population, it is as virulent as ever (as, for instance, the small-pox among the Indians)—but by the selection of a race less subject to attack through the destruction of those that were more so, and the inheritance of the comparative immunity by the children and the grandchildren of the survivors; and how this immunity itself, causing the particular disease to become rare, paves the way to a return of the original fatality; for the mass of such population, both in the present and the immediately preceding generation, not having been exposed to the infection, or but little exposed, has not undergone selection, and so in time the proportion liable to attack, or to fatal attack, gets to be as large as ever. The greater the fatality, especially in the population under marriageable age, the more favorable the condition of the survivors; and, by the law of heredity, their children should share in the immunity. This explanation of the cause, or of one cause, of the return of pests at intervals no less applies to the diminution of the efficacy of remedies, and of preventive means, such as vaccination. When Jenner introduced vaccination, the small-pox in Europe and European colonies must have lost somewhat of its primitive intensity by the vigorous weeding out of the more susceptible through many generations. Upon the residue, vaccination was almost complete protection, and, being generally practised, small-pox consequently became rare. Selection thus ceasing to operate, a population arises which has not been exposed to the contagion, and of which a considerable proportion, under the common law of atavism, comes to be very much in the condition of a people invaded for the first time by the disease. To these, as we might expect, vaccination would prove a less safeguard than to their progenitors three or four generations before.

Mr. Bentham is a veteran systematic botanist of the highest rank and widest knowledge. He had not, so far as we know, touched upon questions of origination in the ante-Darwinian era. The dozen of presidential addresses delivered at anniversary meetings of the Linnean Society, from his assumption of the chair in the year 1862 down to the current year—each devoted to some topic of interest—and his recent "Memoir on Compositæ," summing up the general results of a revision of an order to which a full tenth of all higher plants belong, furnish apt examples both of cautious criticism, conditional assent (as becomes the inaugurator of the quantification of the predicate), and of fruitful application of the new views to various problems concerning the classification and geographical distribution of plants. In his hands the hypothesis is turned at once to practical use as an instrument of investigation, as a means of interrogating Nature. In the result, no doubt seems to be left upon the author's mind that the existing species of plants are the result of the differentiation of previous species, or at least that the derivative hypothesis is to be adopted as that which offers the most natural, if not the only, explanation of the problems concerned. Similar conclusions reached in this country, from a study of the relations of its present flora with that which in earlier ages occupied the arctic zone, might also be referred to. (*See* preceding article.)

An excellent instance of the way in which the derivative hypothesis is practically applied in these days, by a zoölogist, is before us in Prof. Flower's modest and admirable paper on the Ungulata, or hoofed animals, and their geological history. We refer to it here, not so much for the conclusions it reaches or suggests, as to commend the clearness and the impartiality of the handling, and the sobriety and moderation of the deductions. Confining himself "within the region of the known, it is shown that, at least in one group of animals, the facts which we have as yet acquired point to the former existence of various intermediate forms, so numerous that they go far to discredit the view of the sudden introduction of new species. . . . The modern forms are placed along lines which converge toward a common centre." The gaps between the existing forms of the odd-toed group of ungulates (of which horses, rhinoceroses, and tapirs, are the principal representatives) are mostly bridged over by palæontology, and somewhat the same may be said of the even-toed group, to which the ruminants and the porcine genus belong. "Moreover, the lines of both groups to a certain extent approximate, but, within the limits of our knowledge, they do not meet. . . . Was the order according to which the introduction of new forms seems to have taken place since the Eocene then entirely changed, or did it continue as far back as the period when these lines would have been gradually fused in a common centre?"

Facts like these, which suggest grave diversification under long lapse of time, are well supplemented by those which essentially demonstrate a slighter diversification of many species over a wide range of space; whether into species or races depends partly upon how the naturalist uses these terms, partly upon the extent of the observations, or luck in getting together intermediate forms. The researches of Prof. Baird upon the birds of this continent afford a good illustration. A great number of our birds which have been, and must needs have been, regarded as very distinct species, each mainly with its own geographical area, are found to mingle their characters along bordering lines; and the same kinds of differences (of coloration, form, or other) are found to prevail through the species of each region, thus impressing upon them a geographical facies. Upon a submergence of the continent, reducing these several regions to islands sufficiently separated, these forms would be unquestioned species.

Considerations such as these, of which a few specimens have now been adduced (not general speculations, as the unscientific are apt to suppose), and trials of the new views, to see how far they will explain the problems or collocate the facts they are severally dealing with, are what have mainly influenced working naturalists in the direction of the provisional acceptance of the derivative hypothesis. They leave to polemical speculators the fruitless discussion of the question whether all species came from one or two, or more; they are trying to grasp the thing by the near, not by the farther end, and to ascertain, first of all, whether it is probable or provable that present species are descendants of former ones which were like them, but less and less like them the farther back we go.

And it is worth noting that they all seem to be utterly unconscious of wrong-doing. Their repugnance to novel hypotheses is only the natural and healthy one. A change of a wonted line of thought is not made without an effort, nor need be made without adequate occasion. Some courage was required of the man who first swallowed an oyster from its shell; and of most of us the snail would still demand more. As the unaccustomed food proves to be good and satisfying, and also harmless, we may come to like it. That, however, which many good and eminent naturalists find to be healthful and reasonable, and others innocuous, a few still regard as most unreasonable and harmful. At present, we call to mind only two who not only hold to the entire fixity of species as an axiom or a confirmed principle, but also as a dogma, and who maintain, either expressly or implicitly, that the logical antithesis to the creation of species as they are, is not by law (which implies intention), but by chance. A recent book by one of these naturalists, or rather, by a geologist of eminence, the "Story of the Earth and Man," by Dr. Dawson, is now before us. The title is too near that of Guyot's "Earth and Man," with the publication of which popular volume that distinguished physical naturalist commenced his career in this country; and such catch-titles are a sort of trade-mark. As to the nature and merits of Dr. Dawson's work, we have left ourselves space only to say: 1. That it is addressed *ad populum*, which renders it rather the more than less amenable to the criticisms we may be disposed to make upon it. 2. That the author is thoroughly convinced that no species or form deserving the name was ever derived from another, or originated from natural causes; and he maintains this doctrine with earnestness, much variety of argument and illustration, and no small ability; so that he may be taken as a representative of the view exactly opposed to that which is favored by those naturalists whose essays we have been considering—to whom, indeed, he stands in marked contrast in spirit and method, being greatly disposed to argue the question from the remote rather than the near end. 3. And finally, he has a conviction that the evolutionary doctrines of the day are not only untrue, but thoroughly bad and irreligious. This belief, and the natural anxiety with which he contemplates their prevalence, may excuse a certain vehemence and looseness of statement which were better avoided, as where the geologists of the day are said to be "broken up into bands of specialists, little better than scientific banditti, liable to be beaten in detail, and prone to commit outrages on common-sense and good taste which bring their otherwise good cause into disrepute;" and where he despairingly suggests that the prevalence of the doctrines he deprecates "seems to indicate that the accumulated facts of our age have gone altogether beyond its capacity for generalization, and, but for the vigor which one sees everywhere, might be taken as an indication that the human mind has fallen into a state of senility."

This is droll reading, when one considers that the "evolutionist" is the only sort of naturalist who has much occasion to employ his "capacity for generalization" upon "the accumulated facts" in their bearing upon the problem of the origin of species; since the "special creationist," who maintains that they were supernaturally originated just as they are, by the very terms of his doctrine places them out of the reach of scientific explanation. Again, when one reflects upon the new impetus which the derivative hypothesis has given to systematic natural history, and reads the declaration of a master in this department (the President of the Linnean Society) that Mr. Darwin "has in this nineteenth century brought about as great a revolution in the philosophic study of organic Nature as that which was effected in the previous century by the immortal Swede," it sounds oddly to hear from Dr. Dawson that "it obliterates the fine perception of differences from the mind of the naturalist, . . . destroys the possibility of a philosophical classification, reducing all things to a mere series, and leads to a rapid decay in systematic zoölogy and botany, which is already very manifest among the disciples of Spencer and Darwin in England." So, also, "it removes from the study of Nature the ideas of final cause and purpose" —a sentence which reads curiously in the light of Darwin's special investigations, such as those upon the climbing of plants, the agency of insects in the fertilization of blossoms, and the like, which have brought back teleology to natural science, wedded to morphology and already fruitful of discoveries.

The difficulty with Dr. Dawson here is (and it need not be underrated) that apparently he cannot as yet believe an adaptation, act, or result, to be purposed the apparatus of which is perfected or evolved in the course of Nature—a common but a crude state of mind on the part of those who believe that there is any originating purpose in the universe, and one which, we are sure, Dr. Dawson does not share as respects the material world until he reaches the organic kingdoms, and there, possibly, because he sees man at the head of them—of them, while above them. However that may be, the position which Dr. Dawson chooses to occupy is not left uncertain. After concluding, substantially, that those "evolutionists" who exclude design from Nature. thereby exclude theism, which nobody will deny, he proceeds (on page 348) to give his opinion that the "evolutionism which professes to have a creator somewhere behind it. . . . is practically atheistic," and, "if possible, more unphilosophical than that which professes to set out from absolute and eternal nonentity," etc.

There are some sentences which might lead one to suppose that Dr. Dawson himself admitted of an evolution "with a creator somewhere behind it." He offers it (page 320) as a permissible alternative that even man "has been created mediately by the operation of forces also concerned in the production of other animals;" concedes that a just theory "does not even exclude evolution or derivation, to a certain extent" (page 341); and that "a modern man of science" may safely hold "that all things have been produced by the Supreme Creative Will, acting either directly or through the agency of the forces and materials of his own production." Well, if this be so, why denounce the modern man of science so severely upon the other page merely for accepting the permission? At first sight, it might be thought that our author is exposing himself in one paragraph to a share of the condemnation which he deals out in the other. But the permitted views are nowhere adopted as his own; the evolution is elsewhere restricted within specific limits; and as to "mediate creation," although we cannot divine what is here meant by the term, there is reason to think it does not imply that the several species of a genus were mediately created, in a natural way, through the supernatural creation of a remote common ancestor. So that his own judgment in the matter is probably more correctly gathered from the extract above referred to and other similar deliverances, such as that in which he warns those who "endeavor to steer a middle course, and to maintain that the Creator has proceeded by way of evolution," that "the bare, hard logic of Spencer, the greatest English authority on evolution, leaves no place for this compromise, and shows that the theory, carried out to its legitimate consequences, excludes the knowledge of a Creator and the possibility of his work."

Now, this is a dangerous line to take. Those defenders of the faith are more zealous than wise who must needs fire away in their catapults the very bastions of the citadel, in the defense of outposts that have become untenable. It has been and always will be possible to take an atheistic view of Nature, but far more reasonable from science and philosophy only to take a theistic view. Voltaire's saying here holds true: that if there were no God known, it would be necessary to invent one. It is the best, if not the only, hypothesis for the explanation of the facts. Whether the philosophy of Herbert Spencer (which is not to our liking) is here fairly presented, we have little occasion and no time to consider. In this regard, the close of his article No. 12 in the *Contemporary Review* shows, at least, his expectation of the entire permanence of our ideas of cause, origin, and religion, and predicts the futility of the expectation that the "religion of humanity" will be the religion of the future, or "can ever more than temporarily shut out the thought of a Power, of which humanity is but a small and fugitive product, which was in its course of ever-changing manifestation before humanity was, and will continue through other manifestations when humanity has ceased to be." If, on the one hand, the philosophy of the unknowable of the Infinite may be held in a merely quasi-theistic or even atheistic way, were not its ablest expounders and defenders Hamilton and Dean Mansel? One would suppose that Dr. Dawson might discern at least as much of a divine foundation to Nature as Herbert Spencer and Matthew Arnold; might recognize in this power that "something not ourselves that makes" for *order* as well as "for righteousness," and which he fitly terms supreme creative will; and, resting in this, endure with more complacency and faith the inevitable prevalence of evolutionary views which he is powerless to hinder. Although he cannot arrest the stream, he might do something toward keeping it in safe channels.

We wished to say something about the way in which scientific men, worthy of the name, hold hypotheses and theories, using them for the purpose of investigation and the collocation of facts, yielding or withholding assent in degrees or provisionally, according to the amount of verification or likelihood, or holding it long in suspense; which is quite in contrast to that of amateurs and general speculators (not that we reckon Dr. Dawson in this class), whose assent or denial seldom waits, or endures qualification. With them it must on all occasions be yea or nay only, according to the letter of the Scriptural injunction, and whatsoever is *less* than this, or between the two, cometh of evil.

VII.

EVOLUTION AND THEOLOGY.[1]

(THE NATION, *January* 15, 1874.)

THE attitude of theologians toward doctrines of evolution, from the nebular hypothesis down to "Darwinism," is no less worthy of consideration, and hardly less diverse, than that of naturalists. But the topic, if pursued far, leads to questions too wide and deep for our handling here, except incidentally, in the brief notice which it falls in our way to take of the Rev. George Henslow's recent volume on "The Theory of Evolution of Living Things." This treatise is on the side of evolution, "considered as illustrative of the wisdom and beneficence of the Almighty." It

[1] "The Theory of Evolution of Living Things, and the Application of the Principles of Evolution to Religion, considered as illustrative of the 'Wisdom and Beneficence of the Almighty.' By the Rev. George Henslow, M. A., F. L. S., F. G. S., etc." New York: Macmillan & Co. 1873. 12mo, pp. 220.

"Systematic Theology. By Charles Hodge, D. D., Professor in the Theological Seminary, Princeton, New Jersey. Vol. ii. (Part II, Anthropology.") New York: Charles Scribner & Co. 1872.

"Religion and Science: A Series of Sunday Lectures on the Relation of Natural and Revealed Religion, or the Truths revealed in Nature and Scripture. By Joseph Le Conte, Professor of Geology and Natural History in the University of California." New York: D. Appleton & Co. 1874. 12mo, pp. 324.

was submitted for and received one of the Actonian prizes recently awarded by the Royal Institution of Great Britain. We gather that the staple of a part of it is worked up anew from some earlier discourses of the author upon "Genesis and Geology," "Science and Scripture not antagonistic," etc.

In coupling with it a chapter of the second volume of Dr. Hodge's "Systematic Theology (Part II., Anthropology)," we call attention to a recent essay, by an able and veteran writer, on the other side of the question. As the two fairly enough represent the extremes of Christian thought upon the subject, it is convenient to review them in connection. Theologians have a short and easy, if not wholly satisfactory, way of refuting scientific doctrines which they object to, by pitting the authority or opinion of one *savant* against another. Already, amid the currents and eddies of modern opinion, the *savants* may enjoy the same advantage at the expense of the divines—we mean, of course, on the scientific arena; for the mutual refutation of conflicting theologians on their own ground is no novelty. It is not by way of offset, however, that these divergent or contradictory views are here referred to, but only as an illustration of the fact that the divines are by no means all arrayed upon one side of the question in hand. And indeed, in the present transition period, until some one goes much deeper into the heart of the subject, as respects the relations of modern science to the foundations of religious belief, than either of these writers has done, it is as well that the weight of opinion should be distributed, even if only according to prepossessions, rather than that the whole stress should bear upon a single point, and that perhaps the authority of an interpretation of Scripture. A consensus of opinion upon Dr. Hodge's ground, for instance (although better guarded than that of Dr. Dawson), if it were still possible, would—to say the least—probably not at all help to reconcile science and religion. Therefore, it is not to be regretted that the diversities of view among accredited theologians and theological naturalists are about as wide and as equably distributed between the extremes (and we may add that the views themselves are quite as hypothetical) as those which prevail among the various naturalists and natural philosophers of the day.

As a theologian, Mr. Henslow doubtless is not to be compared with the veteran professor at Princeton. On the other hand, he has the advantage of being a naturalist, and the son of a naturalist, as well as a clergyman: consequently he feels the full force of an array of facts in nature, and of the natural inferences from them, which the theological professor, from his Biblical standpoint, and on his implicit assumption that the Old Testament must needs teach true science, can hardly be expected to appreciate. Accordingly, a naturalist would be apt to say of Dr. Hodge's exposition of "theories of the universe" and kindred topics—and in no captious spirit—that whether right or wrong on particular points, he is not often right or wrong in the way of a man of science.

Probably from the lack of familiarity with prevalent ideas and their history, the theologians are apt to suppose that scientific men of the present day are taking up theories of evolution in pure wantonness or mere superfluity of naughtiness; that it would have been quite possible, as well as more proper, to leave all such matters alone. *Quieta non movere* is doubtless a wise rule upon such subjects, so long as it is fairly applicable. But the time for its application in respect to questions of the origin and relations of existing species has gone by. To ignore them is to imitate the foolish bird that seeks security by hiding its head in the sand. Moreover, the naturalists did not force these questions upon the world; but the world they study forced them upon the naturalists. How these questions of derivation came naturally and inevitably to be revived, how the cumulative probability that the existing are derived from preëxisting forms impressed itself upon the minds of many naturalists and thinkers, Mr. Henslow has briefly explained in the introduction and illustrated in the succeeding chapters of the first part of his book. Science, he declares, has been compelled to take up the hypothesis of the evolution of living things as better explaining all the phenomena. In his opinion, it has become "infinitely more probable that all living and extinct beings have been developed or evolved by natural laws of generation from preëxisting forms, than that they, with all their innumerable races and varieties, should owe their existences severally to Creative fiats." This doctrine, which even Dr. Hodge allows may possibly be held in a theistic sense, and which, as we suppose, is so held or viewed by a great proportion of the naturalists of our day, Mr. Henslow maintains is fully compatible with dogmatic as well as natural theology;

that it explains moral anomalies, and accounts for the mixture of good and evil in the world, as well as for the merely relative perfection of things; and, finally, that "the whole scheme which God has framed for man's existence, from the first that was created to all eternity, collapses if the great law of evolution be suppressed." The second part of his book is occupied with a development of this line of argument. By this doctrine of evolution he does not mean the Darwinian hypothesis, although he accepts and includes this, looking upon natural selection as playing an important though not an unlimited part. He would be an evolutionist with Mivart and Owen and Argyll, even if he had not the *vera causa* which Darwin contributed to help him on. And, on rising to man, he takes ground with Wallace, saying:

"I would wish to state distinctly that I do not at present see any evidence for believing in a gradual development of man from the lower animals by ordinary natural laws; that is, without some special interference, or, if it be preferred, some exceptional conditions which have thereby separated him from all other creatures, and placed him decidedly in advance of them all. On the other hand, it would be absurd to regard him as totally severed from them. It is the great degree of difference I would insist upon, bodily, mental, and spiritual, which precludes the idea of his having been evolved by exactly the same processes, and with the same limitations, as, for example, the horse from the palæotherium."

In illustrating this view, he reproduces Wallace's well-known points, and adds one or two of his own. We need not follow up his lines of argument. The essay, indeed, adds nothing material to the discussion of evolution, although it states one side of the case moderately well, as far as it goes.

Dr. Hodge approaches the subject from the side of systematic theology, and considers it mainly in its bearing upon the origin and original state of man. Under each head he first lays down "the Scriptural doctrine," and then discusses "anti-Scriptural theories," which latter, under the first head, are the heathen doctrine of spontaneous generation, the modern doctrine of spontaneous generation, theories of development, specially that of Darwin, the atheistic character of the theory, etc. Although he admits "that there is a theistic and an atheistic form of the nebular hypothesis as to the origin of the universe, so there may be a theistic interpretation of the Darwinian theory," yet he contends that "the system is thoroughly atheistic," notwithstanding that the author "expressly acknowledges the existence of God." Curiously enough, the atheistic form of evolutionary hypotheses, or what he takes for such, is the only one which Dr. Hodge cares to examine. Even the "Reign of Law" theory, Owen's "purposive route of development and change . . . . by virtue of inherent tendencies thereto," as well as other expositions of the general doctrine on a theistic basis, are barely mentioned without a word of comment, except, perhaps, a general "protest against the arraying of probabilities against the teachings of Scripture."

Now, all former experience shows that it is neither safe nor wise to pronounce a whole system "thoroughly atheistic" which it is conceded may be held theistically, and which is likely to be largely held, if not to prevail, on scientific grounds. It may be well to remember that, "of the two great minds of the seventeenth century, Newton and Leibnitz, both profoundly religious as well as philosophical, one produced the theory of gravitation, the other objected to that theory that it was subversive of natural religion; also that the nebular hypothesis—a natural consequence of the theory of gravitation and of the subsequent progress of physical and astronomical discovery—has been denounced as atheistical even down to our day." It has now outlived anathema.

It is undeniable that Mr. Darwin lays himself open to this kind of attack. The propounder of natural selection might be expected to make the most of the principle, and to overwork the law of parsimony in its behalf. And a system in which exquisite adaptation of means to ends, complicated interdependences, and orderly sequences, appear as results instead of being introduced as factors, and in which special design is ignored in the particulars, must needs be obnoxious, unless guarded as we suppose Mr. Darwin might have guarded his ground if he had chosen to do so. Our own opinion, after long consideration, is, that Mr. Darwin has no atheistical intent; and that, as respects the test question of design in Nature, his view may be made clear to the theological mind by likening it to that of the "believer in general but not in particular Providence." There is no need to cull passages in support of this interpretation from his various works while the author—the most candid of men—retains through all the editions of the "Origin of Species" the two mottoes from Whewell and Bishop Butler.[1]

The gist of the matter lies in the answer that should be rendered to the questions—1. Do order and useful-working collocation, pervading a system throughout all its parts, prove design? and, 2. Is such evidence negatived or invalidated by the probability that these particular collocations belong to lineal series of such in time, and diversified in the course of Nature—grown up, so to say, step by step? We do not use the terms "adaptation," "arrangement of means to ends," and the like, because they beg the question in stating it.

Finally, ought not theologians to consider whether they have not already, in principle, conceded to the geologists and physicists all that they are asked to concede to the evolutionists; whether, indeed, the main natural theological difficulties which attend the doctrine of evolution—serious as they may be—are not virtually contained in the admission that there is a system of Nature with fixed laws. This, at least, we may say, that, under a system in which so much is done "by the establishment of general laws," it is

[1] "But with regard to the material world, we can at least go so far as this—we can perceive that events are brought about, not by insulated interpositions of divine power, exerted in each particular case, but by the establishment of general laws."—*Whewell's Bridgewater Treatise.*

"The only distinct meaning of the word 'natural' is *stated*, *fixed*, or *settled;* since what is natural as much requires and presupposes an intelligent agent to render it so—i. e., to effect it continually or at stated times—as what is supernatural or miraculous does to effect it for once."—*Butler's Analogy.*

legitimate for any one to prove, if he can, that any particular thing in the natural world is so done; and it is the proper business of scientific men to push their enquiries in this direction.

It is beside the point for Dr. Hodge to object that, "from the nature of the case, what concerns the origin of things cannot be known except by a supernatural revelation;" that "science has to do with the facts and laws of Nature: here the question concerns the origin of such facts." For the very object of the evolutionists, and of Mr. Darwin in particular, is to remove these subjects from the category of origination, and to bring them under the domain of science by treating them as questions about how things go on, not how they began. Whether the succession of living forms on the earth is or is not among the facts and laws of Nature, is the very matter in controversy.

Moreover, adds Dr. Hodge, it has been conceded that in this matter "proofs, in the proper sense of the word, are not to be had; we are beyond the region of demonstration, and have only probabilities to consider." Wherefore "Christians have a right to protest against the arraying of *probabilities* against the clear teachings of Scripture." The word is italicized, as if to intimate that probabilities have no claims which a theologian is bound to respect. As to arraying them against Scripture, there is nothing whatever in the essay referred to that justifies the statement. Indeed, no occasion offered; for the writer was discussing evolution in its relations to theism, not to Biblical theology, and probably would not be disposed to intermix arguments so different in kind as those from natural science and those from revelation. To pursue each independently, according to its own method, and then to compare the results, is thought to be the better mode of proceeding. The weighing of probabilities we had regarded as a proper exercise of the mind preparatory to forming an opinion. Probabilities, hypotheses, and even surmises, whatever they may be worth, are just what, as it seems to us, theologians ought not to be foremost in decrying, particularly those who deal with the reconciliation of science with Scripture, Genesis with geology, and the like. As soon as they go beyond the literal statements even of the English text, and enter into the details of the subject, they find ample occasion and display a special aptitude for producing and using them, not always with very satisfactory results. It is not, perhaps, for us to suggest that the theological army in the past has been too much encumbered with *impedimenta* for effective aggression in the conflict against atheistic tendencies in modern science; and that in resisting attack it has endeavored to hold too much ground, so wasting strength in the obstinate defense of positions which have become unimportant as well as untenable. Some of the arguments, as well as the guns, which well served a former generation, need to be replaced by others of longer range and greater penetration.

If the theologians are slow to discern the signs and exigencies of the times, the religious philosophical naturalists must be looked to. Since the above remarks were written, Prof. Le Conte's "Religion and Science," just issued, has come to our hands. It is a series of nineteen Sunday lectures on the relation of natural and revealed religion, prepared in the first instance for a Bible-class of young men, his pupils in the University of South Carolina, repeated to similar classes at the University of California, and finally delivered to a larger and general audience. They are printed, the preface states, from a *verbatim* report, with only verbal alterations and corrections of some redundancies consequent upon extemporaneous delivery. They are not, we find, lectures on science under a religious aspect, but discourses upon Christian theology and its foundations from a scientific layman's point of view, with illustrations from his own lines of study. As the headings show, they cover, or, more correctly speaking, range over, almost the whole field of theological thought, beginning with the personality of Deity as revealed in Nature, the spiritual nature and attributes of Deity, and the incarnation; discussing by the way the general relations of theology to science, man, and his place in Nature; and ending with a discussion of predestination and free-will, and of prayer in relation to invariable law—all in a volume of three hundred and twenty-four duodecimo pages! And yet the author remarks that many important subjects have been omitted because he felt unable to present them in a satisfactory manner from a scientific point of view. We note, indeed, that one or two topics which would naturally come in his way—such, especially, as the relation of evolution to the human race—are somewhat conspicuously absent. That most of the momentous subjects which he takes up are treated discursively, and not exhaustively, is all the better for his readers.

What they and we most want to know is, how these serious matters are viewed by an honest, enlightened, and devout scientific man. To solve the mysteries of the universe, as the French lady required a philosopher to explain his new system, "*dans un mot*," is beyond rational expectation.

All that we have time and need to say of this little book upon great subjects relates to its spirit and to the view it takes of evolution. Its theology is wholly orthodox; its tone devotional, charitable, and hopeful; its confidence in religious truth, as taught both in Nature and revelation, complete; the illustrations often happy, but often too rhetorical; the science, as might be expected from this author, unimpeachable as regards matters of fact, discreet as to matters of opinion. The argument from design in the first lecture brings up the subject of the introduction of species. Of this, considered "as a question of history, there is no witness on the stand except geology."

"The present condition of geological evidence is undoubtedly in favor of some degree of suddenness—is against infinite gradations. The evidence may be meagre . . . . but whether meagre or not, it is all the evidence we have. . . . Now, the evidence of geology to-day is, that species seem to come in suddenly and in full perfection, remain substantially unchanged during the term of their existence, and pass away in full perfection. Other species take their place apparently by substitution, not by transmutation. But you will ask me, 'Do you, then, reject the doctrine of evolution? Do you accept the creation of species *directly* and without secondary agencies and processes?' I answer, No! Science knows nothing of phenomena which do not take place by secondary causes and processes. She does not deny such occurrence, for true Science is not dogmatic, and she knows full well that, tracing up the phenomena from cause to cause, we must somewhere reach the more direct agency of a First Cause. . . . It is evident that, however species were introduced, whether suddenly or gradually, it is the duty of Science ever to strive to understand the means and processes by which species originated. . . . Now, of the various conceivable secondary causes and processes, by some of which we must believe species originated, by far the most probable is certainly that of evolution from other species."

[We might interpose the remark that the witness on the stand, if subjected to cross-examination by a biologist, might be made to give a good deal of testimony in favor of transmutation rather than substitution.]

After referring to different ideas as to the cause or mode of evolution, he concludes that it can make no difference, so far as the argument of design in Nature is concerned, whether there be evolution or not, or whether, in the case of evolution, the change be paroxysmal or uniform. We may infer even that he accepts the idea that "physical and chemical forces are changed into vital force, and *vice versa*." Physicists incline more readily to this than physiologists; and if what is called vital force be a force in the physicists' sense, then it is almost certainly so. But the illustration on page 275 touches this point only seemingly. It really concerns only the storing and the using of physical force in a living organism. If, for want of a special expression, we continue to use the term vital force to designate that intangible something which directs and governs the accumulation and expenditure of physical force in organisms, then there is as yet no proof and little likelihood that this is correlate with physical force.

"A few words upon the first chapter of Genesis and the Mosaic cosmogony, and I am done," says Prof. Le Conte, and so are we:

"It might be expected by many that, after speaking of schemes of reconciliation, I should give mine also. My Christian friends, these schemes of reconciliation become daily more and more distasteful to me. I have used them in times past; but now the deliberate construction of such schemes seems to me almost like trifling with the words of Scripture and the teachings of Nature. They seem to me almost irreverent, and quite foreign to the true, humble, liberal spirit of Christianity; they are so evidently artificial, so evidently mere ingenious human devices. It seems to me that if we will only regard the two books in the philosophical spirit which I have endeavored to describe, and then simply wait and possess our souls in patience, the questions in dispute will soon adjust themselves as other similar questions have already done."

VIII.

WHAT IS DARWINISM?[1]

(The Nation, *May* 28, 1874.)

The question which Dr. Hodge asks he promptly and decisively answers: "What is Darwinism? it is atheism."

Leaving aside all subsidiary and incidental matters, let us consider—1. What the Darwinian doctrine is, and 2. How it is proved to be atheistic. Dr. Hodge's own statement of it cannot be very much bettered:

"His [Darwin's] work on the 'Origin of Species' does not purport to be philosophical. In this aspect it is very different from the cognate works of Mr. Spencer. Darwin does not speculate on the origin of the universe, on the nature of matter or of force. He is simply a naturalist, a careful and laborious observer, skillful in his descriptions, and singularly candid in dealing with the difficulties in the way of his peculiar doctrine. He set before himself a single problem—namely, How are the fauna

[1] "What is Darwinism? By Charles Hodge, Princeton, N. J." New York: Scribner, Armstrong & Co. 1874.

"The Doctrine of Evolution. By Alexander Winchell, LL. D., etc." New York: Harper & Brothers. 1874.

"Darwinism and Design; or, Creation by Evolution. By George St. Clair." London: Hodder & Stoughton. 1873.

"Westminster Sermons. By the Rev. Charles Kingsley, F. L. S., F. G. S., Canon of Westminster, etc." London and New York: Macmillan & Co. 1874.

and flora of our earth to be accounted for? . . . To account for the existence of matter and life, Mr. Darwin admits a Creator. This is done explicitly and repeatedly. . . . He assumes the efficiency of physical causes, *showing no disposition to resolve them into mind-force or into the efficiency of the First Cause.* . . . He assumes, also, the existence of life in the form of one or more primordial germs. . . . How all living things on earth, including the endless variety of plants and all the diversity of animals, . . . have descended from the primordial animalcule, he thinks, may be accounted for by the operation of the following natural laws, viz.: First, the law of Heredity, or that by which like begets like—the offspring are like the parent. Second, the law of Variation; that is, while the offspring are in all essential characteristics like their immediate progenitor, they nevertheless vary more or less within narrow limits from their parent and from each other. Some of these variations are indifferent, some deteriorations, some improvements—that is, such as enable the plant or animal to exercise its functions to greater advantage. Third, the law of Over-Production. All plants and animals tend to increase in a geometrical ratio, and therefore tend to overrun enormously the means of support. If all the seeds of a plant, all the spawn of a fish, were to arrive at maturity, in a very short time the world could not contain them. Hence, of necessity, arises a struggle for life. Only a few of the myriads born can possibly live. Fourth, here comes in the law of Natural Selection, or the Survival of the Fittest; that is, if any individual of a given species of plant or animal happens to have a slight deviation from the normal type favorable to its success in the struggle for life, it will survive. This variation, by the law of heredity, will be transmitted to its offspring, and by them again to theirs. Soon these favored ones gain the ascendency, and the less favored perish, and the modification becomes established in the species. After a time, another and another of such favorable variations occur, with like results. Thus, very gradually, great changes of structure are introduced, and not only species, but genera, families, and orders, in the vegetable and animal world, are produced" (pp. 26–29).

Now, the truth or the probability of Darwin's hypothesis is not here the question, but only its congruity or incongruity with theism. We need take only one exception to this abstract of it, but that is an important one for the present investigation. It is to the sentence which we have italicized in the earlier part of Dr. Hodge's own statement of what Darwinism is. With it begins our inquiry as to how he proves the doctrine to be atheistic.

First, if we rightly apprehend it, a suggestion of atheism is infused into the premises in a negative form: Mr. Darwin shows no disposition to resolve the efficiency of physical causes into the efficiency of the First Cause. Next (on page 48) comes the positive charge that "Mr. Darwin, although himself a theist," maintains that "the contrivances manifested in the organs of plants and animals . . . . are not due to the continued coöperation and control of the divine mind, nor to the original purpose of God in the constitution of the universe." As to the negative statement, it might suffice to recall Dr. Hodge's truthful remark that Darwin "is simply a naturalist," and that "his work on the origin of species does not purport to be philosophical." In physical and physiological treatises, the most religious men rarely think it necessary to postulate the First Cause, nor are they misjudged by the omission. But surely Mr. Darwin does show the disposition which our author denies him, not only by implication in many instances, but most explicitly where one would naturally look for it, namely—at the close of the volume in question: "To my mind, it accords better with what we know of the laws impressed on matter by the Creator," etc. If that does not refer the efficiency of physical causes to the First Cause, what form of words could do so? The positive charge appears to be equally gratuitous. In both Dr. Hodge must have overlooked the beginning as well as the end of the volume which he judges so hardly. Just as mathematicians and physicists, in their systems, are wont to postulate the fundamental and undeniable truths they are concerned with, or what they take for such and require to be taken for granted, so Mr. Darwin postulates, upon the first page of his notable work, and in the words of Whewell and Bishop Butler: 1. The establishment by divine power of general laws, according to which, rather than by insulated interpositions in each particular case, events are brought about in the material world; and 2. That by the word "natural" is meant "stated, fixed, or settled," by this same power, "since what is natural as much requires and presupposes an intelligent agent to render it so—i. e., to effect it continually or at stated times—as what is supernatural or miraculous does to effect it for once."[1] So when Mr. Darwin makes such large and free use of "natural as antithetical to supernatural" causes, we are left in no doubt as to the ultimate source which he refers them to. Rather let us say there ought to be no doubt, unless there are other grounds for it to rest upon.

Such ground there must be, or seem to be, to justify or excuse a veteran divine and scholar like Dr. Hodge in his deduction of pure atheism from a system

[1] These two postulate-mottoes are quoted in full in a previous article, in No. 446 of the *Nation* (page 259 of the present volume).

produced by a confessed theist, and based, as we have seen, upon thoroughly orthodox fundamental conceptions. Even if we may not hope to reconcile the difference between the theologian and the naturalist, it may be well to ascertain where their real divergence begins, or ought to begin, and what it amounts to. Seemingly, it is in their proximate, not in their ultimate, principles, as Dr. Hodge insists when he declares that the whole drift of Darwinism is to prove that everything "may be accounted for by the blind operation of natural causes, without any intention, purpose, or coöperation of God" (page 64). "Why don't he say," cries the theologian, "that the complicated organs of plants and animals are the product of the divine intelligence? If God made them, it makes no difference, so far as the question of design is concerned, how he made them, whether at once or by process of evolution" (page 58). But, as we have seen, Mr. Darwin does say that, and he over and over implies it when he refers the production of species "to secondary causes," and likens their origination to the origination of individuals; species being series of individuals with greater difference. It is not for the theologian to object that the power which made individual men and other animals, and all the differences which the races of mankind exhibit, through secondary causes, could not have originated congeries of more or less greatly differing individuals through the same causes.

Clearly, then, the difference between the theologian and the naturalist is not fundamental, and evolution may be as profoundly and as particularly theistic as it is increasingly probable. The taint of atheism which, in Dr. Hodge's view, leavens the whole lump, is not inherent in the original grain of Darwinism—in the principles posited—but has somehow been introduced in the subsequent treatment. Possibly, when found, it may be eliminated. Perhaps there is mutual misapprehension growing out of some ambiguity in the use of terms. "Without any intention, purpose, or coöperation of God." These are sweeping and effectual words. How came they to be applied to natural selection by a divine who professes that God ordained whatsoever cometh to pass? In this wise: "The point to be proved is, that it is the distinctive doctrine of Mr. Darwin that species owe their origin—1. Not to the original intention of the divine mind; 2. Not to special acts of creation calling new forms into existence at certain epochs; 3. Not to the constant and everywhere operative efficiency of God guiding physical causes in the production of intended effects; but 4. To the gradual accumulation of *unintended* variations of structure and instinct securing some advantage to their subjects" (page 52). Then Dr. Hodge adduces "Darwin's own testimony," to the purport that natural selection denotes the totality of natural causes and their interactions, physical and physiological, reproduction, variation, birth, struggle, extinction—in short, all that is going on in Nature; that the variations which in this interplay are picked out for survival are *not intentionally guided;* that "nothing can be more hopeless than the attempt to explain this similarity of pattern in members of the same class by utility or the doctrine of final causes" (which Dr.

Hodge takes to be the denial of any such thing as final causes); and that the interactions and processes going on which constitute natural selection may suffice to account for the present diversity of animals and plants (primordial organisms being postulated and time enough given) with all their structures and adaptations—that is, to account for them scientifically, as science accounts for other things.

A good deal may be made of this, but does it sustain the indictment? Moreover, the counts of the indictment may be demurred to. It seems to us that only one of the three points which Darwin is said to deny is really opposed to the fourth, which he is said to maintain, except as concerns the perhaps ambiguous word *unintended*. Otherwise, the origin of species through the gradual accumulation of variations—i. e., by the addition of a series of small differences—is surely not incongruous with their origin through "the original intention of the divine mind" or through "the constant and everywhere operative efficiency of God." One or both of these Mr. Darwin (being, as Dr. Hodge says, a theist) must needs hold to in some form or other; wherefore he may be presumed to hold the fourth proposition in such wise as not really to contradict the first or the third. The proper antithesis is with the second proposition only, and the issue comes to this: Have the multitudinous forms of living creatures, past and present, been produced by as many special and independent acts of creation at very numerous epochs? Or have they originated under causes as natural as reproduction and birth, and no more so, by the variation and change of preceding into succeeding species?

Those who accept the latter alternative are evolutionists. And Dr. Hodge fairly allows that their views, although clearly wrong, may be genuinely theistic. Surely they need not become the less so by the discovery or by the conjecture of natural operations through which this diversification and continued adaptation of species to conditions is brought about. Now, Mr. Darwin thinks—and by this he is distinguished from most evolutionists—that he can assign actual natural causes, adequate to the production of the present out of the preceding state of the animal and vegetable world, and so on backward—thus uniting, not indeed the beginning but the far past with the present in one coherent system of Nature. But in assigning actual natural causes and processes, and applying them to the explanation of the whole case, Mr. Darwin assumes the obligation of maintaining their general sufficiency—a task from which the numerous advocates and acceptors of evolution on the general concurrence of probabilities and its usefulness as a working hypothesis (with or without much conception of the manner how) are happily free. Having hit upon a *modus operandi* which all who understand it admit will explain something, and many that it will explain very much, it is to be expected that Mr. Darwin will make the most of it. Doubtless he is far from pretending to know all the causes and operations at work; he has already added some and restricted the range of others; he probably looks for additions to their number and new illustrations of their efficiency;

but he is bound to expect them all to fall within the category of what he calls natural selection (a most expansible principle), or to be congruous with it—that is, that they shall be natural causes. Also—and this is the critical point—he is bound to maintain their sufficiency without *intervention.*

Here, at length, we reach the essential difference between Darwin, as we understand him, and Dr. Hodge. The terms which Darwin sometimes uses, and doubtless some of the ideas they represent, are not such as we should adopt or like to defend; and we may say once for all—aside though it be from the present issue—that, in our opinion, the adequacy of the assigned causes to the explanation of the phenomena has not been made out. But we do not understand him to deny "purpose, intention, or the coöperation of God" in Nature. This would be as gratuitous as unphilosophical, not to say unscientific. When he speaks of this or that particular or phase in the course of events or the procession of organic forms as not intended, he seems to mean not specially and disjunctively intended and not brought about by intervention. Purpose in the whole, as we suppose, is not denied but implied. And when one considers how, under whatever view of the case, the designed and the contingent lie inextricably commingled in this world of ours, past man's disentanglement, and into what metaphysical dilemmas the attempt at unraveling them leads, we cannot greatly blame the naturalist for relegating such problems to the philosopher and the theologian. If charitable, these will place the most favorable construction upon attempts to extend and unify the operation of known secondary causes, this being the proper business of the naturalist and physicist; if wise, they will be careful not to predicate or suggest the absence of intention from what comes about by degrees through the continuous operation of physical causes, even in the organic world, lest, in their endeavor to retain a probable excess of supernaturalism in that realm of Nature, they cut away the grounds for recognizing it at all in inorganic Nature, and so fall into the same condemnation that some of them award to the Darwinian.

Moreover, it is not certain that Mr. Darwin would very much better his case, Dr. Hodge being judge, if he did propound some theory of the *nexus* of divine causation and natural laws, or even if he explicitly adopted the one or the other of the views which he is charged with rejecting. Either way he might meet a procrustean fate; and, although a saving amount of theism might remain, he would not be sound or comfortable. For, if he predicates "the constant and everywhere operative efficiency of God," he may "lapse into the same doctrine" that the Duke of Argyll and Sir John Herschel "seem inclined to," the latter of whom is blamed for thinking "it but reasonable to regard the force of gravitation as the direct or indirect result of a consciousness or will existing somewhere," and the former for regarding "it unphilosophical 'to think or speak as if the forces of Nature were either independent of or even separate from the Creator's power'" (page 24): while if he falls back upon an "original intention of the divine mind," endowing matter with forces which he foresaw and intended should produce such results as these contrivances in Nature, he is told (pages 44–46) that this banishes God from the world, and is inconsistent with obvious facts. And that because of its implying that "He never *interferes* to guide the operation of physical causes." We italicize the word, for *interference* proves to be the keynote of Dr. Hodge's system. Interference with a divinely ordained physical Nature for the accomplishment of natural results! An unorthodox friend has just imparted to us, with much misgiving and solicitude lest he should be thought irreverent, his tentative hypothesis, which is, that even the Creator may be conceived to have improved with time and experience! Never before was this theory so plainly and barely put before us. We were obliged to say that, in principle and by implication, it was not wholly original.

But in such matters, which are far too high for us, no one is justly to be held responsible for the conclusions which another may draw from his principles or assumptions. Dr. Hodge's particular view should be gathered from his own statement of it:

"In the external world there is always and everywhere indisputable evidence of the activity of two kinds of force, the one physical, the other mental. The physical belongs to matter, and is due to the properties with which it has been endowed; the other is the everywhere present and ever-acting mind of God. To the latter are to be referred all the manifestations of design in Nature, and the ordering of events in Providence. This doctrine does not ignore the efficiency of second causes; it simply asserts that God overrules and controls them. Thus the Psalmist says: 'I am fearfully and wonderfully made. My substance was not hid from Thee when I was made in secret, and curiously wrought (or embroidered) in the lower parts of the earth. . . . God makes the grass to grow, and herbs for the children of men.' He sends rain, frost, and snow. He controls the winds and the waves. He determines the casting of the lot, the flight of an arrow, and the falling of a sparrow" (pages 43, 44).

Far be it from us to object to this mode of conceiving divine causation, although, like the two other theistic conceptions referred to, it has its difficulties, and perhaps the difficulties of both. But, if we understand it, it draws an unusually hard and fast line between causation in organic and inorganic Nature, seems to look for no manifestation of design in the latter except as "God overrules and controls" second causes, and, finally, refers to this overruling and controlling (rather than to a normal action through endowment) all embryonic development, the growth of vegetables, and the like. He even adds, without break or distinction, the sending of rain, frost, and snow, the flight of an arrow, and the falling of a sparrow. Somehow we must have misconceived the bearing of the statement; but so it stands as one of "the three ways," and the right way, of "accounting for contrivances in Nature;" the other two being—1. Their reference to the blind operation of natural causes; and, 2. That they were foreseen and purposed by God, who endowed matter with forces which he foresaw and intended should produce such results, but never *interferes* to guide their operation.

In animadverting upon this latter view, Dr. Hodge brings forward an argument against evolution, with the examination of which our remarks must close:

"Paley, indeed, says that if the construction of a watch be an undeniable evidence of design, it would be a still more wonderful manifestation of skill if a watch could be made to produce other watches, and, it may be added, not only other watches, but all kinds of timepieces, in endless variety. So it has been asked, If a man can make a telescope, why cannot God make a telescope which produces others like itself? This is simply asking whether matter can be made to do the work of mind. The idea involves a contradiction. For a telescope to make a telescope supposes it to select copper and zinc in due proportions, and fuse them into brass; to fashion that brass into inter-entering tubes; to collect and combine the requisite materials for the different kinds of glass needed; to melt them, grind, fashion, and polish them, adjust their densities, focal distances, etc., etc. A man who can believe that brass can do all this might as well believe in God" (pp. 45, 46).

If Dr. Hodge's meaning is, that matter unconstructed cannot do the work of mind, he misses the point altogether; for original construction by an intelligent mind is given in the premises. If he means that the machine cannot originate the power that operates it, this is conceded by all except believers in perpetual motion, and it equally misses the point; for the operating power is given in the case of the watch, and implied in that of the reproductive telescope. But if he means that matter cannot be made to do the work of mind in constructions, machines, or organisms, he is surely wrong. "*Solvitur ambulando,*" *vel scribendo;* he confuted his argument in the act of writing the sentence. That is just what machines and organisms are for; and a consistent Christian theist should maintain that it is what all matter is for. Finally, if, as we freely suppose, he means none of these, he must mean (unless we are much mistaken) that organisms originated by the Almighty Creator could not be endowed with the power of producing similar organisms, or slightly dissimilar organisms, without successive interventions. Then he begs the very question in dispute, and that, too, in the face of the primal command, "Be fruitful and multiply," and its consequences in every natural birth. If the actual facts could be ignored, how nicely the parallel would run! "The idea involves a contradiction." For an animal to make an animal, or a plant to make a plant, supposes it to select carbon, hydrogen, oxygen, and nitrogen, to combine these into cellulose and protoplasm, to join with these some phosphorus, lime, etc., to build them into structures and usefully-adjusted organs. A man who can believe that plants and animals can do this (not, indeed, in the crude way suggested, but in the appointed way) "might as well believe in God." Yes, verily, and so he probably will, in spite of all that atheistical philosophers have to offer, if not harassed and confused by such arguments and statements as these.

There is a long line of gradually-increasing divergence from the ultra-orthodox view of Dr. Hodge through those of such men as Sir William Thomson, Herschel, Argyll, Owen, Mivart, Wallace, and Darwin, down to those of Strauss, Vogt, and Büchner. To strike the line with telling power and good effect, it is necessary to aim at the right place. Excellent as the present volume is in motive and clearly as it shows that Darwinism may bear an atheistic as well as a theistic interpretation, we fear that it will not contribute much to the reconcilement of science and religion.

The length of the analysis of the first book on our list precludes the notices which we intended to take of the three others. They are all the production of men who are both scientific and religious, one of them a celebrated divine and writer unusually versed in natural history. They all look upon theories of evolution either as in the way of being established or as not unlikely to prevail, and they confidently expect to lose thereby no solid ground for theism or religion. Mr. St. Clair, a new writer, in his "Darwinism and Design; or, Creation by Evolution," takes his ground in the following succinct statement of his preface:

"It is being assumed by our scientific guides that the design-argument has been driven out of the field by the doctrine of evolution. It seems to be thought by our theological teachers that the best defense of the faith is to deny evolution *in toto*, and denounce it as anti-Biblical. My volume endeavors to show that, if evolution be true, all is not lost; but, on the contrary, something is gained: the design-argument remains unshaken, and the wisdom and beneficence of God receive new illustration."

Of his closing remark, that, so far as he knows, the subject has never before been handled in the same way for the same purpose, we will only say that the handling strikes us as mainly sensible rather than as substantially novel. He traverses the whole ground of evolution, from that of the solar system to "the origin of moral species." He is clearly a theistic Darwinian without misgiving, and the arguments for that hypothesis and for its religious aspects obtain from him their most favorable presentation, while he combats the *dysteleology* of Häckel, Büchner, etc., not, however, with any remarkable strength.

Dr. Winchell, chancellor of the new university at Syracuse, in his volume just issued upon the "Doctrine of Evolution," adopts it in the abstract as "clearly as the law of universal intelligence under which complex results are brought into existence" (whatever that may mean), accepts it practically for the inorganic world as a geologist should, hesitates as to the organic world, and sums up the arguments for the origin of species by diversification unfavorably for the Darwinians, regarding it mainly from the geological side. As some of our zoölogists and palæontologists may have somewhat to say upon this matter, we leave it for their consideration. We are tempted to develop a point which Dr. Winchell incidentally refers to—viz., how very modern the idea of the independent creation and fixity of species is, and how well the old divines got on without it. Dr. Winchell reminds us that St. Augustine and St. Thomas Aquinas were model evolutionists; and, where authority is deferred to, this should count for something.

Mr. Kingsley's eloquent and suggestive "Westminster Sermons," in which he touches here and there upon many of the topics which evolution brings up, has incorporated into the preface a paper which he read in 1871 to a meeting of London clergy at Sion College, upon certain problems of natural theology as affected by modern theories in science. We may hereafter have occasion to refer to this volume. Meanwhile, perhaps we may usefully conclude this article with two or three short extracts from it:

"The God who satisfies our conscience ought more or less to satisfy our reason also. To teach that was Butler's mission;

and he fulfilled it well. But it is a mission which has to be refulfilled again and again, as human thought changes, and human science develops. For if, in any age or country, the God who seems to be revealed by Nature seems also different from the God who is revealed by the then-popular religion, then that God and the religion which tells of that God will gradually cease to be believed in.

"For the demands of reason—as none knew better than good Bishop Butler—must be and ought to be satisfied. And, therefore, when a popular war arises between the reason of any generation and its theology, then it behooves the ministers of religion to inquire, with all humility and godly fear, on whose side lies the fault; whether the theology which they expound is all that it should be, or whether the reason of those who impugn it is all that it should be."

Pronouncing it to be the duty of the naturalist to find out the how of things, and of the natural theologian to find out the why, Mr. Kingsley continues:

"But if it be said, 'After all, there is no why; the doctrine of evolution, by doing away with the theory of creation, does away with that of final causes,' let us answer boldly, 'Not in the least.' We might accept all that Mr. Darwin, all that Prof. Huxley, all that other most able men have so learnedly and acutely written on physical science, and yet preserve our natural theology on the same basis as that on which Butler and Paley left it. That we should have to develop it I do not deny.

"Let us rather look with calmness, and even with hope and good-will, on these new theories; they surely mark a tendency toward a more, not a less, Scriptural view of Nature.

"Of old it was said by Him, without whom nothing is made, 'My Father worketh hitherto, and I work.' Shall we quarrel with Science if she should show how these words are true? What, in one word, should we have to say but this: 'We know of old that God was so wise that he could make all things; but, behold, he is so much wiser than even that, that he can make all things make themselves?'"

IX.

CHARLES DARWIN: A SKETCH.

(NATURE, *June* 4, 1874, ACCOMPANYING A PORTRAIT.)

Two British naturalists, Robert Brown and Charles Darwin, have, more than any others, impressed their influence upon science in this nineteenth century. Unlike as these men and their works were and are, we may most readily subserve the present purpose in what we are called upon to say of the latter by briefly comparing and contrasting the two.

Robert Brown died sixteen years ago, full of years and scientific honors, and he seems to have finished, several years earlier, all the scientific work that he had undertaken. To the other, Charles Darwin, a fair number of productive years may yet remain, and are earnestly hoped for. Both enjoyed the great advantage of being all their lives long free from exacting professional duties or cares, and so were able in the main to apply themselves to research without distraction and according to their bent. Both, at the beginning of their career, were attached to expeditions of exploration in the southern hemisphere, where they amassed rich stores of observation and materials, and probably struck out, while in the field, some of the best ideas which they subsequently developed. They worked in different fields and upon different methods;

only in a single instance, so far as we know, have they handled the same topic; and in this the more penetrating insight of the younger naturalist into an interesting general problem may be appealed to in justification of a comparison which some will deem presumptuous. Be this as it may, there will probably be little dissent from the opinion that the characteristic trait common to the two is an unrivaled scientific sagacity. In this these two naturalists seem to us, each in his way, preeminent. There is a characteristic likeness, too—underlying much difference—in their admirable manner of dealing with facts closely, and at first hand, without the interposition of the formal laws, vague ideal conceptions, or "glittering generalities" which some philosophical naturalists make large use of.

A likeness may also be discerned in the way in which the works or contributions of predecessors and contemporaries are referred to. The brief historical summaries prefixed to many of Mr. Brown's papers are models of judicial conscientiousness. And Mr. Darwin's evident delight at discovering that some one else has "said his good things before him," or has been on the verge of uttering them, seemingly equals that of making the discovery himself. It reminds one of Goethe's insisting that his views in morphology must have been held before him and must be somewhere on record, so obvious did they appear to him.

Considering the quiet and retired lives led by both these men, and the prominent place they are likely to occupy in the history of science, the contrast between them as to contemporary and popular fame is very remarkable. While Mr. Brown was looked up to with the greatest reverence by all the learned botanists, he was scarcely heard of by any one else; and out of botany he was unknown to science except as the discoverer of the Brownian motion of minute particles, which discovery was promulgated in a privately-printed pamphlet that few have ever seen. Although Mr. Darwin had been for twenty years well and widely known for his "Naturalist's Journal," his works on "Coral Islands," on "Volcanic Islands," and especially for his researches on the Barnacles, it was not till about fifteen years ago that his name became popularly famous. Ever since no scientific name has been so widely spoken. Many others have had hypotheses or systems named after them, but no one else that we know of a department of bibliography. The nature of his latest researches accounts for most of the difference, but not for all. The Origin of Species is a fascinating topic, having interests and connections with every branch of science, natural and moral. The investigation of recondite affinities is very dry and special; its questions, processes, and results alike—although in part generally presentable in the shape of morphology—are mainly, like the higher mathematics, unintelligible except to those who make them a subject of serious study. They are especially so when presented in Mr. Brown's manner. Perhaps no naturalist ever recorded the results of his investigations in fewer words and with greater precision than Robert Brown: certainly no one ever took more pains to state nothing beyond the precise point in question. Indeed, we have sometimes fancied that he preferred to enwrap rather than to explain his meaning; to put it into such a form that, unless you follow Solomon's injunction and dig for the wisdom as for hid treasure, you may hardly apprehend it until you have found it all out for yourself, when you will have the satisfaction of perceiving that Mr. Brown not only knew all about it, but had put it upon record. Very different from this is the way in which Mr. Darwin takes his readers into his confidence, freely displays to them the sources of his information, and the working of his mind, and even shares with them all his doubts and misgivings, while in a clear exposition he sets forth the reasons which have guided him to his conclusions. These you may hesitate or decline to adopt, but you feel sure that they have been presented with perfect fairness; and if you think of arguments against them you may be confident that they have all been duly considered before.

The sagacity which characterizes these two naturalists is seen in their success in finding decisive instances, and their sure insight into the meaning of things. As an instance of the latter on Mr. Darwin's part, and a justification of our venture to compare him with the *facile princeps botanicorum*, we will, in conclusion, allude to the single instance in which they took the same subject in hand. In his papers on the organs and modes of fecundation in *Orchideæ* and *Asclepiadeæ*, Mr. Brown refers more than once to C. K. Sprengel's almost forgotten work, shows how the structure of the flowers in these orders largely requires the agency of insects for their fecundation, and is aware that "in *Asclepiadeæ* . . . . the insect so readily passes from one corolla to another that it not unfrequently visits every flower of the umbel." He must also have contemplated the transport of pollen from plant to plant by wind and insects; and we know from another source that he looked upon Sprengel's ideas as far from fantastic. Yet, instead of taking the single forward step which now seems so obvious, he even hazarded the conjecture that the insect-forms of some orchideous flowers are intended to deter rather than to attract insects. And so the explanation of all these and other extraordinary structures, as well as of the arrangement of blossoms in general, and even the very meaning and need of sexual propagation, were left to be supplied by Mr. Darwin. The aphorism "Nature abhors a vacuum" is a characteristic specimen of the science of the middle ages. The aphorism "Nature abhors close fertilization," and the demonstration of the principle, belong to our age, and to Mr. Darwin. To have originated this, and also the principle of natural selection—the truthfulness and importance of which are evident the moment it is apprehended—and to have applied these principles to the system of Nature in such a manner as to make, within a dozen years, a deeper impression upon natural history than has been made since Linnæus, is ample title for one man's fame.

There is no need of our giving any account or of estimating the importance of such works as the "Origin of Species by means of Natural Selection," the "Variation of Animals and Plants under Domestication," the "Descent of Man, and Selection in relation to Sex," and the "Expression of the Emotions in Man and Animals"—a series to which we may hope other volumes may in due time be added. We would rather, if space permitted, attempt an analysis of the less known, but not less masterly, subsidiary essays, upon the various arrangements for insuring cross-fertilization in flowers, for the climbing of plants, and the like. These, as we have heard, may before long be reprinted in a volume, and supplemented by some long-pending but still unfinished investigations upon the action of *Dionœa* and *Drosera*—a capital subject for Mr. Darwin's handling.

À propos to these papers, which furnish excellent illustrations of it, let us recognize Darwin's great service to natural science in bringing back to it Teleology; so that, instead of Morphology *versus* Teleology, we shall have Morphology wedded to Teleology. To many, no doubt, evolutionary Teleology comes in such a questionable shape as to seem shorn of all its goodness; but they will think better of it in time, when their ideas become adjusted, and they see what an impetus the new doctrines have given to investigation. They are much mistaken who suppose that Darwinism is only of speculative importance, and perhaps transient interest. In its working applications it has proved to be a new power, eminently practical and fruitful.

And here, again, we are bound to note a striking contrast to Mr. Brown, greatly as we revere his memory. He did far less work than was justly to be expected from him. Mr. Darwin not only points out the road, but labors upon it indefatigably and unceasingly. A most commendable *noblesse oblige* assures us that he will go on while strength (would we could add health) remains. The vast amount of such work he has already accomplished might overtax the powers of the strongest. That it could have been done at all under constant infirm health is most wonderful.

X.

INSECTIVOROUS PLANTS.

(THE NATION, *April* 2 and 9, 1874.)

THAT animals should feed upon plants is natural and normal, and the reverse seems impossible. But the adage, "*Natura non agit saltatim*," has its application even here. It is the naturalist, rather than Nature, that draws hard and fast lines everywhere, and marks out abrupt boundaries where she shades off with gradations. However opposite the parts which animals and vegetables play in the economy of the world as the two opposed kingdoms of organic Nature, it is becoming more and more obvious that they are not only two contiguous kingdoms, but are parts of one whole—antithetical and complementary to each other, indeed; but such "thin partitions do the bounds divide" that no definitions yet framed hold good without exception. This is a world of transition in more senses than is commonly thought; and one of the lessons which the philosophical naturalist learns, or has to learn, is, that differences the most wide and real in the main, and the most essential, may nevertheless be here and there connected or bridged over by gradations. There is a limbo filled with organisms which never rise high enough in the scale to be manifestly either animal or plant, unless it may be said of some of them that they are each in turn and neither long. There are undoubted animals which produce the essential material of vegetable fabric, or build up a part of their structure of it, or elaborate the characteristic *leaf-green* which, under solar light, assimilates inorganic into organic matter, the most distinguishing function of vegetation. On the other hand, there are plants—microscopic, indeed, but unquestionable—which move spontaneously and freely around and among animals that are fixed and rooted. And, to come without further parley to the matter in hand, while the majority of animals feed directly upon plants, "for 'tis their nature to," there are plants which turn the tables and feed upon them. Some, being parasitic upon living animals, feed insidiously and furtively; these, although really cases in point, are not so extraordinary, and, as they belong to the lower orders, they are not much regarded, except for the harm they do. There are others, and those of the highest orders, which lure or entrap animals in ways which may well excite our special wonder—all the more so since we are now led to conclude that they not only capture but consume their prey.

As respects the two or three most notable instances, the conclusions which have been reached are among the very recent acquisitions of physiological science. Curiously enough, however, now that they are made out, it appears that they were in good part long ago attained, recorded, and mainly forgotten. The earlier observations and surmises shared the common fate of discoveries made before the time, or by those who were not sagacious enough to bring out their full meaning or importance. Vegetable morphology, dimly apprehended by Linnæus, initiated by Caspar Frederick Wolff, and again, independently in successive generations, by Goethe and by De Candolle, offers a parallel instance. The botanists of Goethe's day could not see any sense, advantage, or practical application, to be made of the proposition that the parts of a blossom answer to leaves; and so the study of homologies had long to wait. Until lately it appeared to be of no consequence whatever (except, perhaps, to the insects) whether Drosera and Sarracenia caught flies or not; and even Dionæa excited only unreflecting wonder as a vegetable anomaly. As if there were real anomalies in Nature, and some one plant possessed extraordinary powers denied to all others, and (as was supposed) of no importance to itself!

That most expert of fly-catchers, Dionæa, of which so much has been written and so little known until lately, came very near revealing its secret to Solander and Ellis a hundred years ago, and doubtless to John Bartram, our botanical pioneer, its probable discoverer, who sent it to Europe. Ellis, in his published letter to Linnæus, with which the history begins, described the structure and action of the living trap correctly; noticed that the irritability which called forth the quick movement closing the trap, entirely resided in the few small bristles of its upper face; that this whole surface was studded with glands, which probably secreted a liquid; and that the trap did not open again when an insect was captured, even upon the death of the captive, although it opened very soon when nothing was caught, or when the irritation was caused by a bit of straw, or any such substance. It was Linnæus who originated the contrary and erroneous statement, which has long prevailed in the books, that the trap reopened when the fatigued captive became quiet, and let it go; as if the plant caught flies in mere play and pastime! Linnæus also omitted all allusion to a secreted liquid—which was justifiable, as Ellis does not state that he had actually seen any; and, if he did see it, quite mistook its use, supposing it to be, like the nectar of flowers, a lure for insects, a bait for the trap. Whereas, in fact, the lure, if there be any, must be an odor (although nothing is perceptible to the human olfactories); for the liquid secreted by the glands never appears until the trap has closed upon some insect, and held it at least for some hours a prisoner. Within twenty-four or forty-eight hours this glairy liquid is abundant, bathing and macerating the body of the perished insect. Its analogue is not the nectar of flowers, but the saliva or the gastric juice!

The observations which compel such an inference are recent, and the substance of them may be briefly stated. The late Rev. Dr. M. A. Curtis (by whose death, two years ago, we lost one of our best botanists, and the master in his especial line, mycology), forty years and more ago resided at Wilmington, North Carolina, in the midst of the only district to which the Dionæa is native; and he published, in 1834, in the first volume of the "Journal of the Boston Society of Natural History," by far the best account of this singular plant which had then appeared. He remarks that "the little prisoner is not crushed and suddenly destroyed, as is sometimes supposed," for he had often liberated "captive flies and spiders, which sped away as fast as fear or joy could hasten them." But he neglected to state, although he must have noticed the fact, that the two sides of the trap, at first concave to the contained insect, at length flatten and close down firmly upon the prey, exerting no inconsiderable pressure, and insuring the death of any soft-bodied insect, if it had not already succumbed to the confinement and salivation. This last Dr. Curtis noticed, and first discerned its import, although he hesitated to pronounce upon its universality. That the captured insects were in some way "made subservient to the nourishment of the plant" had been conjectured from the first. Dr. Curtis "at times [and he might have always at the proper time] found them enveloped in a fluid of mucilaginous consistence, which seems to act as a solvent, the insects being more or less consumed in it." This was verified and the digestive character of the liquid well-nigh demonstrated six or seven years ago by Mr. Canby, of Wilmington, Delaware, who, upon a visit to the sister-town of North Carolina, and afterward at his home, followed up Dr. Curtis's suggestions with some capital observations and experiments. These were published at Philadelphia in the tenth volume of Meehan's *Gardeners' Monthly*, August, 1868; but they do not appear to have attracted the attention which they merited.

The points which Mr. Canby made out are, that

this fluid is always poured out around the captured insect in due time, "if the leaf is in good condition and the prey suitable;" that it comes from the leaf itself, and not from the decomposing insect (for, when the trap caught a plum-curculio, the fluid was poured out while he was still alive, though very weak, and endeavoring, ineffectually, to eat his way out); that bits of raw beef, although sometimes rejected after a while, were generally acted upon in the same manner—i. e., closed down upon tightly, slavered with the liquid, dissolved mainly, and absorbed; so that, in fine, the fluid may well be said to be analogous to the gastric juice of animals, dissolving the prey and rendering it fit for absorption by the leaf. Many leaves remain inactive or slowly die away after one meal; others reopen for a second and perhaps even a third capture, and are at least capable of digesting a second meal.

Before Mr. Canby's experiments had been made, we were aware that a similar series had been made in England by Mr. Darwin, with the same results, and with a small but highly-curious additional one—namely, that the fluid secreted in the trap of Dionæa, like the gastric juice, has an acid reaction. Having begun to mention unpublished results (too long allowed to remain so), it may be well, under the circumstances, to refer to a still more remarkable experiment by the same most sagacious investigator. By a prick with a sharp lancet at a certain point, he has been able to paralyze one-half of the leaf-trap, so that it remained motionless under the stimulus to which the other half responded. Such high and sensitive organization entails corresponding ailments. Mr. Canby tells us that he gave to one of his Dionæa-subjects a fatal dyspepsia by feeding it with cheese; and under Mr. Darwin's hands another suffers from *paraplegia*.

Finally, Dr. Burdon-Sanderson's experiments, detailed at the last meeting of the British Association for the Advancement of Science, show that the same electrical currents are developed upon the closing of the Dionæa-trap as in the contraction of a muscle.

If the Venus's Fly-trap stood alone, it would be doubly marvelous—first, on account of its carnivorous propensities, and then as constituting a real anomaly in organic Nature, to which nothing leads up. Before acquiescing in such a conclusion, the modern naturalist would scrutinize its relatives. Now, the nearest relatives of our vegetable wonder are the sundews.

While Dionæa is as local in habitation as it is singular in structure and habits, the Droseras or sundews are widely diffused over the world and numerous in species. The two whose captivating habits have attracted attention abound in bogs all around the northern hemisphere. That flies are caught by them is a matter of common observation; but this was thought to be purely accidental. They spread out from the root a circle of small leaves, the upper face of which especially is beset and the margin fringed with stout bristles (or what seem to be such, although the structure is more complex), tipped by a secreting gland, which produces, while in vigorous state, a globule of clear liquid like a drop of dew—whence the name, both Greek and English. One expects these seeming dew-drops to be dissipated by the morning sun; but they remain unaffected. A touch shows that the glistening drops are glutinous and extremely tenacious, as flies learn to their cost on alighting, perhaps to sip the tempting liquid, which acts first as a decoy and then like birdlime. A small fly is held so fast, and in its struggles comes in contact with so many of these glutinous globules, that it seldom escapes.

The result is much the same to the insect, whether captured in the trap of Dionæa or stuck fast to the limed bristles of Drosera. As there are various plants upon whose glandular hairs or glutinous surfaces small insects are habitually caught and perish, it might be pure coincidence that the most effectual arrangement of the kind happens to occur in the nearest relatives of Dionæa. Roth, a keen German botanist of the eighteenth century, was the first to detect, or at least to record, some evidence of intention in Drosera, and to compare its action with that of Dionæa, which, through Ellis's account, had shortly before been made known in Europe. He noticed the telling fact that not only the bristles which the unfortunate insect had come in contact with, but also the surrounding rows, before widely spreading, curved inward one by one, although they had not been touched, so as within a few hours to press their glutinous tips likewise against the body of the captive insect—thus doubling or quadrupling the bonds of the victim and (as we may now suspect) the surfaces through which some part of the animal substance may be imbibed. For Roth surmised that both these plants were, in their way, predaceous. He even observed that the disk of the Drosera-leaf itself often became concave and enveloped the prey. These facts, although mentioned now and then in some succeeding works, were generally forgotten, except that of the adhesion of small insects to the leaves of sundews, which must have been observed in every generation. Up to and even within a few years past, if any reference was made to these asserted movements (as by such eminent physiologists as Meyen and Treviranus) it was to discredit them. Not because they are difficult to verify, but because, being naturally thought improbable, it was easier to deny or ignore them. So completely had the knowledge of almost a century ago died out in later years that, when the subject was taken up anew in our days by Mr. Darwin, he had, as we remember, to advertise for it, by sending a "note and query" to the magazines, asking where any account of the fly-catching of the leaves of sundew was recorded.

When Mr. Darwin takes a matter of this sort in hand, he is not likely to leave it where he found it. He not only confirmed all Roth's observations as to the incurving of the bristles toward and upon an insect entangled on any part of the disk of the leaf, but also found that they responded similarly to a bit of muscle or other animal substance, while to any particles of inorganic matter they were nearly indifferent. To minute fragments of carbonate of ammonia, however, they were more responsive. As these remarkable results, attained (as we are able to attest) half a dozen years ago, remained unpublished (being portions of an investigation not yet completed), it would have been hardly proper to mention them, were it not that independent observers were beginning to bring out the same or similar facts. Mrs. Treat, of New Jersey, noticed the habitual infolding of the leaf in the longer-leaved species of sundew (*American Journal of Science* for November, 1871), as was then thought for the first time—Roth's and Withering's observations not having been looked up. In recording this, the next year, in a very little book, entitled "How Plants Behave," the opportunity was taken to mention, in the briefest way, the capital discovery of Mr. Darwin that the leaves of Drosera act differently when different objects are placed upon them, the bristles closing upon a particle of raw meat as upon a living insect, while to a particle of chalk or wood they are nearly inactive. The same facts were independently brought out by Mr. A. W. Bennett at the last year's meeting of the British Association for the Advancement of Science, and have been mentioned in the journals.

If to these statements, which we may certify, were added some far more extraordinary ones, communicated to the French Academy of Science in May last by M. Zeigler, a stranger story of discrimination on the part of sundew-bristles would be told. But it is safer to wait for the report of the committee to which these marvels were referred, and conclude this sufficiently "strange eventful history" with some details of experiments made last summer by Mrs. Treat, of New Jersey, and published in the December number of the *American Naturalist*. It is well to note that Mrs. Treat selects for publication the observations of one particular day in July, when the sundew-leaves were unusually active; for their moods vary with the weather, and also in other unaccountable ways, although in general the sultrier days are the most appetizing:

"At fifteen minutes past ten of the same day I placed bits of raw beef on some of the most vigorous leaves of Drosera longifolia. Ten minutes past twelve, two of the leaves had folded around the beef, hiding it from sight. Half-past eleven of the same day, I placed living flies on the leaves of D. longifolia. At 12° 48′ one of the leaves had folded entirely around its victim, the other leaves had partially folded, and the flies had ceased to struggle. By 2° 30′ four leaves had each folded around a fly. . . . I tried mineral substances—bits of dry chalk, magnesia, and pebbles. In twenty-four hours, neither the leaves nor their bristles had made any move like clasping these articles. I wet a piece of chalk in water, and in less than an hour the bristles were curving about it, but soon unfolded again, leaving the chalk free on the blade of the leaf." Parallel experiments made on D. rotundifolia, with bits of beef and of chalk, gave the same results as to the action of the bristles; while with a piece of raw apple, after eleven hours, "part of the bristles were clasping it, but not so closely as the beef," and in twenty-four hours "nearly all the bristles were curved toward it, but not many of the glands were touching it."

To make such observations is as easy as it is interesting. Throughout the summer one has only to transfer plants of Drosera from the bogs into pots or pans filled with wet moss—if need be, allowing them to become established in the somewhat changed conditions, or even to put out fresh leaves—and to watch their action or expedite it by placing small flies upon the disk of the leaves. The more common round-leaved sundew acts as well as the other by its bristles, and the leaf itself is sometimes almost equally prehensile, although in a different way, infolding the whole border instead of the summit only. Very curious, and even somewhat painful, is the sight when a fly, alighting upon the central dew-tipped bristles, is held as fast as by a spider's web; while the efforts to escape not only entangle the insect more hopelessly as they exhaust its strength, but call into action the surrounding bristles, which, one by one, add to the number of the bonds, each by itself apparently feeble, but in their combination so effectual that the fly may be likened to the sleeping Gulliver made fast in the tiny but multitudinous toils of the Liliputians. Anybody who can believe that such an apparatus was not intended to capture flies might say the same of a spider's web.

Is the intention here to be thought any the less real because there are other species of Drosera which are not so perfectly adapted for fly-catching, owing to the form of their leaves and the partial or total want of coöperation of their scattered bristles? One such species, *D. filiformis*, the thread-leaved sundew, is not uncommon in this country, both north and south of the district that Dionæa locally inhabits. Its leaves are long and thread-shaped, beset throughout with glutinous gland-tipped bristles, but wholly destitute of a blade. Flies, even large ones, and even moths and butterflies, as Mrs. Treat and Mr. Canby affirm (in the *American Naturalist*), get stuck fast to these bristles, whence they seldom escape. Accidental as such captures are, even these thread-shaped leaves respond more or less to the contact, somewhat in the manner of their brethren. In Mr. Canby's recent and simple experiments, made at Mr. Darwin's suggestion, when a small fly alights upon a leaf a little below its slender apex, or when a bit of crushed fly is there affixed, within a few hours the tip of the leaf bends at the point of contact, and curls over or around the body in question; and Mrs. Treat even found that when living flies were pinned at half an inch in distance from the leaves, these in forty minutes had bent their tips perceptibly toward the flies, and in less than two hours reached them! If this be confirmed—and such a statement needs ample confirmation—then it may be suspected that these slender leaves not only incurve after prolonged contact, just as do the leaf-stalks of many climbers, but also make free and independent circular sweeps, in the manner of twining stems and of many tendrils.

Correlated movements like these indicate purpose. When performed by climbing plants, the object and the advantage are obvious. That the apparatus and the actions of Dionæa and Drosera are purposeless and without advantage to the plants themselves, may have been believed in former days, when it was likewise conceived that abortive and functionless organs were specially created "for the sake of symmetry" and to display a plan; but this is not according to the genius of modern science.

In the cases of insecticide next to be considered, such evidence of intent is wanting, but other and circumstantial evidence may be had, sufficient to warrant conviction. Sarracenias have hollow leaves in the form of pitchers or trumpet-shaped tubes, containing water, in which flies and other insects are habitually drowned. They are all natives of the eastern side of

North America, growing in bogs or low ground, so that they cannot be supposed to need the water as such. Indeed, they secrete a part if not all of it. The commonest species, and the only one at the North, which ranges from Newfoundland to Florida, has a broad-mouthed pitcher with an upright lid, into which rain must needs fall more or less. The yellow Sarracenia, with long tubular leaves, called "trumpets" in the Southern States, has an arching or partly upright lid, raised well above the orifice, so that some water may rain in; but a portion is certainly secreted there, and may be seen bedewing the sides and collected at the bottom before the mouth opens. In other species, the orifice is so completely overarched as essentially to prevent the access of water from without. In these tubes, mainly in the water, flies and other insects accumulate, perish, and decompose. Flies thrown into the open-mouthed tube of the yellow Sarracenia, even when free from water, are unable to get out—one hardly sees why, except that they cannot fly directly upward; and microscopic *chevaux-de-frise* of fine, sharp-pointed bristles which line most of the interior, pointing strictly downward, may be a more effectual obstacle to crawling up the sides than one would think possible. On the inside of the lid or hood of the purple Northern species, the bristles are much stronger; but an insect might escape by the front without encountering these. In this species, the pitchers, however, are so well supplied with water that the insects which somehow are most abundantly attracted thither are effectually drowned, and the contents all summer long are in the condition of a rich liquid manure.

That the tubes or pitchers of the Southern species are equally attractive and fatal to flies is well known. Indeed, they are said to be taken into houses and used as fly-traps. There is no perceptible odor to draw insects, except what arises from the decomposition of macerated victims; nor is any kind of lure to be detected at the mouth of the pitcher of the common purple-flowered species. Some incredulity was therefore natural when it was stated by a Carolinian correspondent (Mr. B. F. Grady) that in the long-leaved, yellow-flowered species the lid just above the mouth of the tubular pitcher habitually secretes drops of a sweet and viscid liquid, which attracts flies and apparently intoxicates them, since those that sip it soon become unsteady in gait and mostly fall irretrievably into the well beneath. But upon cultivating plants of this species, obtained for the purpose, the existence of this lure was abundantly verified; and, although we cannot vouch for its inebriating quality, we can no longer regard it as unlikely.

No sooner was it thus ascertained that at least one species of Sarracenia allures flies to their ruin than it began to appear that—just as in the case of Drosera—most of this was a mere revival of obsolete knowledge. The "insect-destroying process" was known and well described sixty years ago, the part played by the sweet exudation indicated, and even the intoxication perhaps hinted at, although evidently little thought of in those ante-temperance days. Dr. James Macbride, of South Carolina—the early associate of Elliott in his "Botany of South Carolina and Georgia," and to whose death, at the age of thirty-three, cutting short a life of remarkable promise, the latter touchingly alludes in the preface to his second volume—sent to Sir James Edward Smith an account of his observations upon this subject, made in 1810 and the following years. This was read to the Linnæan Society in 1815, and published in the twelfth volume of its "Transactions." From this forgotten paper (to which attention has lately been recalled) we cull the following extracts, premising that the observations mostly relate to a third species, *Sarracenia adunca, alias variolaris*, which is said to be the most efficient fly-catcher of the kind:

"If, in the months of May, June, or July, when the leaves of those plants perform their extraordinary functions in the greatest perfection, some of them be removed to a house and fixed in an erect position, it will soon be perceived that flies are attracted by them. These insects immediately approach the fauces of the leaves, and, leaning over their edges, appear to sip with eagerness something from their internal surfaces. In this position they linger; but at length, allured as it would seem by the pleasure of taste, they enter the tubes. The fly which has thus changed its situation will be seen to stand unsteadily; it totters for a few seconds, slips, and falls to the bottom of the tube, where it is either drowned or attempts in vain to ascend against the points of the hairs. The fly seldom takes wing in its fall and escapes. . . . In a house much infested with flies, this entrapment goes on so rapidly that a tube is filled in a few hours, and it becomes necessary to add water, the natural quantity being insufficient to drown the imprisoned insects. The leaves of *S. adunca* and *rubra* [a fourth species] might well be employed as fly-catchers; indeed, I am credibly informed they are in some neighborhoods. The leaves of the *S. flava* [the species to which our foregoing remarks mainly relate], although they are very capacious, and often grow to the height of three feet or more, are never found to contain so many insects as those of the species above mentioned.

"The cause which attracts flies is evidently a sweet, viscid substance resembling honey, secreted by or exuding from the internal surface of the tube. . . . From the margin, where it commences, it does not extend lower than one-fourth of an inch.

"The falling of the insect as soon as it enters the tube is wholly attributable to the downward or inverted position of the hairs of the internal surface of the leaf. At the bottom of a tube split open, the hairs are plainly discernible pointing downward; as the eye ranges upward, they gradually become shorter and attenuated, till at or just below the surface covered by the bait they are no longer perceptible to the naked eye nor to the most delicate touch. It is here that the fly cannot take a hold sufficiently strong to support itself, but falls. The inability of insects to crawl up against the points of the hairs I have often tested in the most satisfactory manner."

From the last paragraph it may be inferred that Dr. Macbride did not suspect any inebriating property in the nectar, and in a closing note there is a conjecture of an impalpable loose powder in *S. flava*, at the place where the fly stands so unsteadily, and from which it is supposed to slide. We incline to take Mr. Grady's view of the case.

The complete oblivion into which this paper and the whole subject had fallen is the more remarkable when it is seen that both are briefly but explicitly referred to in Elliott's book, with which botanists are familiar.

It is not so wonderful that the far earlier allusion to these facts by the younger Bartram should have been overlooked or disregarded. With the genuine love of Nature and fondness for exploration, William Bartram did not inherit the simplicity of his father, the earliest native botanist of this country. Fine writing was his foible; and the preface to his well-known "Travels" (published at Philadelphia in 1791) is its full-blown illustration, sometimes perhaps deserving the epithet which he applies to the palms of Florida—that of pomposity. In this preface he declares that "all the Sarracenias are insect-catchers, and so is the *Drosera rotundifolia.* Whether the insects caught in their leaves, and which dissolve and mix with the fluid, serve for aliment or support to these kind of plants is doubtful," he thinks, but he should be credited with the suggestion. In one sentence he speaks of the quantities of insects which, "being invited down to sip the mellifluous exuvia from the interior surface of the tube, where they inevitably perish," being prevented from returning by the stiff hairs all pointing downward. This, if it refers to the sweet secretion, would place it below, and not, as it is, above the bristly surface, while the liquid below, charged with decomposing insects, is declared in an earlier sentence to be "cool and animating, limpid as the morning dew." Bartram was evidently writing from memory; and it is very doubtful if he ever distinctly recognized the sweet exudation which entices insects.

Why should these plants take to organic food more than others? If we cannot answer the question, we may make a probable step toward it. For plants that are not parasitic, these, especially the sundews, have much less than the ordinary amount of chlorophyll—that is, of the universal leaf-green upon which the formation of organic matter out of inorganic materials depends. These take it instead of making it, to a certain extent.

What is the bearing of these remarkable adaptations and operations upon doctrines of evolution? There seems here to be a field on which the specific creationist, the evolutionist with design, and the necessary evolutionist, may fight out an interesting, if not decisive, "triangular duel."

XI.

INSECTIVOROUS AND CLIMBING PLANTS.

(The Nation, *January* 6 and 13, 1876.)

"Minerals grow; vegetables grow and live; animals grow, live, and feel;" this is the well-worn, not to say out-worn, diagnosis of the three kingdoms by Linnæus. It must be said of it that the agreement indicated in the first couplet is unreal, and that the distinction declared in the second is evanescent. Crystals do not grow at all in the sense that plants and animals grow. On the other hand, if a response to external impressions by special movements is evidence of feeling, vegetables share this endowment with animals; while, if conscious feeling is meant, this can be affirmed only of the higher animals. What appears to remain true is, that the difference is one of successive addition. That the increment in the organic world is of many steps; that in the long series no absolute

[1] "Insectivorous Plants. By Charles Darwin, M. A., F. R. S." With Illustrations. London: John Murray. 1875. Pp. 462. New York; D. Appleton & Co.

"The Movements and Habits of Climbing Plants. By Charles Darwin, M. A., F. R. S., etc." Second Edition, revised, with Illustrations. London: John Murray. 1875. Pp. 208. New York: D. Appleton & Co.

lines separate, or have always separated, organisms which barely respond to impressions from those which more actively and variously respond, and even from those that consciously so respond—this, as we all know, is what the author of the works before us has undertaken to demonstrate. Without reference here either to that part of the series with which man is connected, and in some sense or other forms a part of, or to that lower *limbo* where the two organic kingdoms apparently merge—or whence, in evolutionary phrase, they have emerged—Mr. Darwin, in the present volumes, directs our attention to the behavior of the highest plants alone. He shows that some (and he might add that all) of them execute movements for their own advantage, and that some capture and digest living prey. When plants are seen to move and to devour, what faculties are left that are distinctively animal?

As to insectivorous or otherwise carnivorous plants, we have so recently here discussed this subject—before it attained to all this new popularity—that a brief account of Mr. Darwin's investigation may suffice.[1] It

[1] The *Nation*, Nos. 457, 458, 1874. It was in these somewhat light and desultory, but substantially serious, articles that some account of Mr. Darwin's observations upon the digestive powers of *Drosera* and *Dionæa* first appeared; in fact, their leading motive was to make sufficient reference to his then unpublished discoveries to guard against expected or possible claims to priority. Dr. Burdon-Sanderson's lecture, and the report in *Nature*, which first made them known in England, appeared later.

A mistake on our part in the reading of a somewhat ambiguous sentence in a letter led to the remark, at the close of the first of those articles (p. 295), that the leaf-trap of *Dionæa* had been paralyzed on one side in consequence of a dexterous puncture. What was communicated really related to *Drosera*.

is full of interest as a physiological research, and is a model of its kind, as well for the simplicity and directness of the means employed as for the clearness with which the results are brought out—results which any one may verify now that the way to them is pointed out, and which, surprising as they are, lose half their wonder in the ease and sureness with which they seem to have been reached.

Rather more than half the volume is devoted to one subject, the round-leaved sundew (*Drosera rotundifolia*), a rather common plant in the northern temperate zone. That flies stick fast to its leaves, being limed by the tenacious seeming dew-drops which stud its upper face and margins, had long been noticed in Europe and in this country. We have heard hunters and explorers in our Northern woods refer with satisfaction to the fate which in this way often befalls one of their plagues, the black fly of early summer. And it was known to some observant botanists in the last century, although forgotten or discredited in this, that an insect caught on the viscid glands it has happened to alight upon is soon fixed by many more—not merely in consequence of its struggles, but by the spontaneous incurvation of the stalks of surrounding and untouched glands; and even the body of the leaf had been observed to incurve or become cup-shaped so as partly to involve the captive insect.

Mr. Darwin's peculiar investigations not only confirm all this, but add greater wonders. They relate to the sensitiveness of these *tentacles*, as he prefers to call them, and the mode in which it is manifested; their power of absorption; their astonishing discernment of the presence of animal or other soluble azotized matter, even in quantities so minute as to rival the spectroscope—that most exquisite instrument of modern research—in delicacy; and, finally, they establish the fact of a true digestion, in all essential respects similar to that of the stomach of animals.

First as to sensitiveness and movement. Sensitiveness is manifested by movement or change of form in response to an external impression. The sensitiveness in the sundew is all in the gland which surmounts the tentacle. To incite movement or other action, it is necessary that the gland itself should be reached. Anything laid on the surface of the viscid drop, the spherule of clear, glairy liquid which it secretes, produces no effect unless it sinks through to the gland; or unless the substance is soluble and reaches it in solution, which, in the case of certain substances, has the same effect. But the glands themselves do not move, nor does any neighboring portion of the tentacle. The outer and longer tentacles bend inward (toward the centre of the leaf) promptly, when the gland is irritated or stimulated, sweeping through an arc of 180° or less, or more—the quickness and the extent of the inflection depending, in equally vigorous leaves, upon the amount of irritation or stimulation, and also upon its kind. A tentacle with a particle of raw meat on its gland sometimes visibly begins to bend in ten seconds, becomes strongly incurved in five minutes, and its tip reaches the centre of the leaf in half an hour; but this is a case of extreme rapidity. A particle of cinder, chalk, or sand, will also incite the bending, if actually brought in contact with the gland, not merely resting on the drop; but the inflection is then much less pronounced and more transient. Even a bit of thin human hair, only $\frac{1}{8000}$ of an inch in length, weighing only the $\frac{1}{78740}$ of a grain, and largely supported by the viscid secretion, suffices to induce movement; but, on the other hand, one or two momentary, although rude, touches with a hard object produce no effect, although a repeated touch or the slightest pressure, such as that of a gnat's foot, prolonged for a short time, causes bending. The seat of the movement is wholly or nearly confined to a portion of the lower part of the tentacle, above the base, where local irritation produces not the slighest effect. The movement takes place only in response to some impression made upon its own gland at the distant extremity, or upon other glands far more remote. For if one of these members suffers irritation the others sympathize with it. Very noteworthy is the correlation between the central tentacles, upon which an insect is most likely to alight, and these external and larger ones, which, in proportion to their distance from the centre, take the larger share in the movement. The shorter central ones do not move at all when a bit of meat, or a crushed fly, or a particle of a salt of ammonia, or the like, is placed upon them; but they transmit their excitation across the leaf to the surrounding tentacles on all sides; and they, although absolutely untouched, as they successively receive the mysterious impulse, bend strongly inward, just as they do when their own glands are excited. Whenever a tentacle bends in obedience to an impulse from its own gland, the movement is always toward the centre of the leaf; and this also takes place, as we have seen, when an exciting object is lodged at the centre. But when the object is placed upon either half of the leaf, the impulse radiating thence causes all the surrounding untouched tentacles to bend with precision toward the point of excitement, even the central tentacles, which are motionless when themselves charged, now responding to the call. The inflection which follows mechanical irritation or the presence of any inorganic or insoluble body is transient; that which follows the application of organic matter lasts longer, more or less, according to its nature and the amount; but sooner or later the tentacles resume their former position, their glands glisten anew with fresh secretion, and they are ready to act again.

As to how the impulse is originated and propagated, and how the movements are made, comparatively simple as the structure is, we know as little as we do of the nature of nervous impulse and muscular motion. But two things Mr. Darwin has wellnigh made out, both of them by means and observations so simple and direct as to command our confidence, although they are contrary to the prevalent teaching. First, the transmission is through the ordinary cellular tissue, and not through what are called the fibrous or vascular bundles. Second, the movement is a vital one, and is effected by contraction on the side toward which the bending takes place, rather than by turgescent tension of the opposite side. The tentacle is pulled over rather than pushed over. So far all accords with muscular action.

The operation of this fly-catching apparatus, in any case, is plain. If the insect alights upon the disk of the leaf, the viscid secretion holds it fast—at least, an ordinary fly is unable to escape—its struggles only increase the number of glands involved and the amount of excitement; this is telegraphed to the surrounding and successively longer tentacles, which bend over in succession, so that within ten to thirty hours, if the leaf is active and the fly large enough, every one of the glands (on the average, nearly two hundred in number) will be found applied to the body of the insect. If the insect is small, and the lodgment toward one side, only the neighboring tentacles may take part in the capture. If two or three of the strong marginal tentacles are first encountered, their prompt inflection carries the intruder to the centre, and presses it down upon the glands which thickly pave the floor; these notify all the surrounding tentacles of the capture, that they may share the spoil, and the fate of that victim is even as of the first. A bit of meat or a crushed insect is treated in the same way.

This language implies that the animal matter is in some way or other discerned by the tentacles, and is appropriated. Formerly there was only a presumption of this, on the general ground that such an organization could hardly be purposeless. Yet, while such expressions were natural, if not unavoidable, they generally were used by those familiar with the facts in a half-serious, half-metaphorical sense. Thanks to Mr. Darwin's investigations, they may now be used in simplicity and seriousness.

That the glands secrete the glairy liquid of the drop is evident, not only from its nature, but from its persistence through a whole day's exposure to a summer sun, as also from its renewal after it has been removed, dried up, or absorbed. That they absorb as well as secrete, and that the whole tentacle may be profoundly affected thereby, are proved by the different effects, in kind and degree, which follow the application of different substances. Drops of rain-water, like single momentary touches of a solid body, produce no effect, as indeed they could be of no advantage; but a little carbonate of ammonia in the water, or an infusion of meat, not only causes inflection, but promptly manifests its action upon the contents of the cells of which the tentacle is constructed. These cells are sufficiently transparent to be viewed under the microscope without dissection or other interference; and the change which takes place in the fluid contents of these cells, when the gland above has been acted upon, is often visible through a weak lens, or sometimes even by the naked eye, although higher powers are required to discern what actually takes place. This change, which Mr. Darwin discovered, and turns to much account in his researches, he terms "aggregation of the protoplasm." When untouched and quiescent, the contents appear as an homogeneous purple fluid. When the gland is acted upon, minute purple particles appear, suspended in the now colorless or almost colorless fluid; and this change appears first in the cells next the gland, and then in those next beneath, traveling down the whole length of the tentacle. When the action is slight, this appearance does not last long; the particles of "aggregated protoplasm" redissolve, the process of redissolution traveling upward from the base of the tentacle to the gland in a reverse direction to that of the aggregation. Whenever the action is more prolonged or intense, as when a bit of meat or crushed fly, or a fitting solution, is left upon the gland, the aggregation proceeds further, so that the whole protoplasm of each cell condenses into one or two masses, or into a single mass which will often separate into two, which afterward reunite; indeed, they incessantly change their forms and positions, being never at rest, although their movements are rather slow. In appearance and movements they are very like amœbæ and the white corpuscles of the blood. Their motion, along with the streaming movement of rotation in the layer of white granular protoplasm that flows along the walls of the cell, under the high powers of the microscope "presents a wonderful scene of vital activity." This continues while the tentacle is inflected or the gland fed by animal matter, but vanishes by dissolution when the work is over and the tentacle straightens. That absorption takes place, and matter is conveyed from cell to cell, is well made out, especially by the experiments with carbonate of ammonia. Nevertheless, this aggregation is not dependent upon absorption, for it equally occurs from mechanical irritation of the gland, and always accompanies inflection, however caused, though it may take place without it. This is also apparent from the astonishingly minute quantity of certain substances which suffices to produce sensible inflection and aggregation—such, for instance, as the

$\frac{1}{20000000}$ or even the $\frac{1}{30000000}$ of a grain of phosphate or nitrate of ammonia!

By varied experiments it was found that the nitrate of ammonia was more powerful than the carbonate, and the phosphate more powerful than the nitrate, this result being intelligible from the difference in the amount of nitrogen in the first two salts, and from the presence of phosphorus in the third. There is nothing surprising in the absorption of such extremely dilute solutions by a gland. As our author remarks: "All physiologists admit that the roots of plants absorb the salts of ammonia brought to them by the rain; and fourteen gallons of rain-water [i. e., early rain-water] contain a grain of ammonia; therefore, only a little more than twice as much as in the weakest solution employed by me. The fact which appears truly wonderful is that the $\frac{1}{20000000}$ of a grain of the phosphate of ammonia, including less than $\frac{1}{30000000}$ of efficient matter [if the water of crystallization is deducted], when absorbed by a gland, should induce some change in it which leads to a motor impulse being transmitted down the whole length of the tentacle, causing its basal part to bend, often through an angle of 180°." But odoriferous particles which act upon the nerves of animals must be infinitely smaller, and by these a dog a quarter of a mile to the leeward of a deer perceives his presence by some change in the olfactory nerves transmitted through them to the brain.

When Mr. Darwin obtained these results, fourteen years ago, he could claim for *Drosera* a power and delicacy in the detection of minute quantities of a substance far beyond the resources of the most skillful chemist; but in a foot-note he admits that "now the spectroscope has altogether beaten *Drosera;* for, according to Bunsen and Kirchhoff, probably less than the $\frac{1}{200000000}$ of a grain of sodium can be thus detected."

Finally, that this highly-sensitive and active living organism absorbs, will not be doubted when it is proved to digest, that is, to dissolve otherwise insoluble animal matter by the aid of special secretions. That it does this is now past doubting. In the first place, when the glands are excited they pour forth an increased amount of the ropy secretion. This occurs directly when a bit of meat is laid upon the central glands; and the influence which they transmit to the long-stalked marginal glands causes them, while incurving their tentacles, to secrete more copiously long before they have themselves touched anything. The primary fluid, secreted without excitation, does not of itself digest. But the secretion under excitement changes in Nature and becomes acid. So, according to Schiff, mechanical irritation excites the glands of the stomach to secrete an acid. In both this acid appears to be necessary to, but of itself insufficient for, digestion. The requisite solvent, a kind of ferment called *pepsin*, which acts only in the presence of the acid, is poured forth by the glands of the stomach only after they have absorbed certain soluble nutritive substances of the food; then this pepsin promptly dissolves muscle, fibrine, coagulated albumen, cartilage, and the like. Similarly it appears that *Drosera*-glands, after irritation by particles of glass, did not act upon little cubes of albumen. But when moistened with saliva, or replaced by bits of roast-meat or gelatine, or even cartilage, which supply some soluble *peptone*-matter to initiate the process, these substances are promptly acted upon, and dissolved or digested; whence it is inferred that the analogy with the stomach holds good throughout, and that a ferment similar to pepsin is poured out under the stimulus of some soluble animal matter. But the direct evidence of this is furnished only by the related carnivorous plant, *Dionœa*, from which the secretions, poured out when digestion is about to begin, may be collected in quantity sufficient for chemical examination. In short, the experiments show "that there is a remarkable accordance in the power of digestion between the gastric juice of animals, with its pepsin and hydrochloric acid, and the secretion of *Drosera*, with its ferment and acid belonging to the acetic series. We can, therefore, hardly doubt that the ferment in both cases is closely similar, if not identically the same. That a plant and an animal should pour forth the same, or nearly the same, complex secretion, adapted for the same purpose of digestion, is a new and wonderful fact in physiology."

There are one or two other species of sundew—one of them almost as common in Europe and North America as the ordinary round-leaved species—which act in the same way, except that, having their leaves longer in proportion to their breadth, their sides never curl inward, but they are much disposed to aid the action of their tentacles by incurving the tip of the leaf, as if to grasp the morsel. There are many others, with variously less efficient and less advantageously arranged insectivorous apparatus, which, in the language of the new science, may be either on the way to acquire something better, or of losing what they may have had, while now adapting themselves to a proper vegetable life. There is one member of the family (*Drosophyllum Lusitanicum*), an almost shrubby plant, which grows on dry and sunny hills in Portugal and Morocco—which the villagers call "the fly-catcher," and hang up in their cottages for the purpose—the glandular tentacles of which have wholly lost their powers of movement, if they ever had any, but which still secrete, digest, and absorb, being roused to great activity by the contact of any animal matter. A friend of ours once remarked that it was fearful to contemplate the amount of soul that could be called forth in a dog by the sight of a piece of meat. Equally wonderful is the avidity for animal food manifested by these vegetable tentacles, that can "only stand and wait" for it.

Only a brief chapter is devoted to *Dionœa* of North Carolina, the Venus's fly-trap, albeit, "from the rapidity and force of its movements, one of the most wonderful in the world." It is of the same family as the sundew; but the action is transferred from tentacles on the leaf to the body of the leaf itself, which is transformed into a spring-trap, closing with a sudden movement over the alighted insect. No secretion is provided beforehand either for allurement or detention; but after the captive is secured, microscopic glands within the surface of the leaf pour out an abundant gastric juice to digest it. Mrs. Glass's classical directions in the cook-book, "first catch your hare," are implicitly followed.

Avoiding here all repetition or recapitulation of our former narrative, suffice it now to mention two interesting recent additions to our knowledge, for which we are indebted to Mr. Darwin. One is a research, the other an inspiration. It is mainly his investigations which have shown that the glairy liquid, which is poured upon and macerates the captured insect, accomplishes a true digestion; that, like the gastric juice of animals, it contains both a free acid and pepsin or its analogue, these two together dissolving albumen, meat, and the like. The other point relates to the significance of a peculiarity in the process of capture. When the trap suddenly incloses an insect which has betrayed its presence by touching one of the internal sensitive bristles, the closure is at first incomplete. For the sides approach in an arching way, surrounding a considerable cavity, and the marginal spine-like bristles merely intercross their tips, leaving intervening spaces through which one may look into the cavity beneath. A good idea may be had of it by bringing the two palms near together to represent the sides of the trap, and loosely interlocking the fingers to represent the marginal bristles or bars. After remaining some time in this position the closure is made complete by the margins coming into full contact, and the sides finally flattening down so as to press firmly upon the insect within; the secretion excited by contact is now poured out, and digestion begins. Why these two stages? Why should time be lost by this preliminary and incomplete closing? The query probably was never distinctly raised before, no one noticing anything here that needed explanation. Darwinian teleology, however, raises questions like this, and Mr. Darwin not only propounded the riddle but solved it. The object of the partial closing is to permit small insects to escape through the meshes, detaining only those plump enough to be worth the trouble of digesting. For naturally only one insect is caught at a time, and digestion is a slow business with Dionæas, as with anacondas, requiring ordinarily a fortnight. It is not worth while to undertake it with a gnat when larger game may be had. To test this happy conjecture, Mr. Canby was asked, on visiting the Dionæas in their native habitat, to collect early in the season a good series of leaves in the act of digesting naturally-caught insects. Upon opening them it was found that ten out of fourteen were engaged upon relatively large prey, and of the remaining four three had insects as large as ants, and one a rather small fly.

"There be land-rats and water-rats" in this carnivorous sundew family. *Aldrovanda*, of the warmer parts of Europe and of India, is an aquatic plant, with bladdery leaves, which were supposed to be useful in rendering the herbage buoyant in water. But it has recently been found that the bladder is composed of two lobes, like the trap of its relative *Dionæa*, or the valves of a mussel-shell; that these open when the plant is in an active state, are provided with some sensitive bristles within, and when these are touched close with a quick movement. These water-traps are manifestly adapted for catching living creatures; and the few incomplete investigations that have already been made render it highly probable that they appropriate their prey for nourishment; whether by digestion or by mere absorption of decomposing animal matter, is uncertain. It is certainly most remarkable that this family of plants, wherever met with, and under the most diverse conditions and modes of life, should always in some way or other be predaceous and carnivorous.

If it be not only surprising but somewhat confounding to our classifications that a whole group of plants should subsist partly by digesting animal matter and partly in the normal way of decomposing carbonic acid and producing the basis of animal matter, we have, as Mr. Darwin remarks, a counterpart anomaly in the animal kingdom. While some plants have stomachs, some animals have roots. "The rhizocephalous crustaceans do not feed like other animals by their mouths, for they are destitute of an alimentary canal, but they live by absorbing through root-like processes the juices of the animals on which they are parasitic."

To a naturalist of our day, imbued with those ideas of the solidarity of organic Nature which such facts as those we have been considering suggest, the greatest anomaly of all would be that they are really anomalous or unique. Reasonably supposing, therefore, that the sundew did not stand alone, Mr. Darwin turned his attention to other groups of plants; and, first, to the bladderworts, which have no near kinship with the sundews, but, like the aquatic representative of that family, are provided with bladdery sacs, under water. In the common species of *Utricularia* or bladderwort, these little sacs, hanging from submerged leaves or branches, have their orifice closed by a lid which opens inwardly—a veritable trap-door. It had been noticed in England and France that they contained minute crustacean animals. Early in the summer of 1874, Mr. Darwin ascertained the mechanism for their capture and the great success with which it is used. But before his account was written out, Prof. Cohn published an excellent paper on the subject in Germany; and Mrs. Treat, of Vineland, New Jersey, a still earlier one in this country—in the *New York Tribune* in the autumn of 1874. Of the latter, Mr. Darwin remarks that she "has been more successful than any other observer in witnessing the actual entrance of these minute creatures." They never come out, but soon perish in their prison, which receives a continued succession of victims, but little, if any, fresh air to the contained water. The action of the trap is purely mechanical, without evident irritability in the opening or shutting. There is no evidence nor much likelihood of proper digestion; indeed, Mr. Darwin found evidence to the contrary. But the more or less decomposed and dissolved animal matter is doubtless absorbed into the plant; for the whole interior of the sac is lined with peculiar, elongated and four-armed very thin-walled processes, which contain active protoplasm, and which were proved by experiment to "have the power of absorbing matter from weak solutions of certain salts of ammonia and urea, and from a putrid infusion of raw meat."

Although the bladderworts "prey on garbage," their terrestrial relatives "live cleanly," as nobler plants should do, and have a good and true digestion. *Pinguicula*, or butterwort, is the representative of this family upon land. It gets both its Latin and its English name from the fatty or greasy appearance of the upper face of its broad leaves; and this appearance is due to a dense coat or pile of short-stalked glands, which secrete a colorless and extremely viscid liquid. By this small flies, or whatever may alight or fall upon the leaf, are held fast. These waifs might be useless or even injurious to the plant. Probably Mr. Darwin was the first to ask whether they might be of advantage. He certainly was the first to show that they probably are so. The evidence from experiment, shortly summed up, is, that insects alive or dead, and also other nitrogenous bodies, excite these glands to increased secretion; the secretion then becomes acid, and acquires the power of dissolving solid animal substances—that is, the power of digestion in the manner of *Drosera* and *Dionæa*. And the stalks of their glands under the microscope give the same ocular evidence of absorption. The leaves of the butterwort are apt to have their margins folded inward, like a rim or hem. Taking young and vigorous leaves to which hardly anything had yet adhered, and of which the margins were still flat, Mr. Darwin set within one margin a row of small flies. Fifteen hours afterward this edge was neatly turned inward, partly covering the row of flies, and the surrounding glands were secreting copiously. The other edge remained flat and unaltered. Then he stuck a fly to the middle of the leaf just below its tip, and soon both margins infolded, so as to clasp the object. Many other and varied experiments yielded similar results. Even pollen, which would not rarely be lodged upon these leaves, as it falls from surrounding wind-fertilized plants, also small seeds, excited the same action, and showed signs of being acted upon. "We may therefore conclude," with Mr. Darwin, "that *Pinguicula vulgaris*, with its small roots, is not only supported to a large extent by the extraordinary number of insects which it habitually captures, but likewise draws some nourishment from the pollen, leaves, and seeds, of other plants which often adhere to its leaves. It is, therefore, partly a vegetable as well as an animal feeder."

What is now to be thought of the ordinary glandular hairs which render the surface of many and the most various plants extremely viscid? Their number is legion. The Chinese primrose of common garden and house culture is no extraordinary instance; but Mr. Francis Darwin, counting those on a small space measured by the micrometer, estimated them at 65,371 to the square inch of foliage, taking in both surfaces of the leaf, or two or three millions on a moderate-sized specimen of this small herb. Glands of this sort were loosely regarded as organs for excretion, without much consideration of the question whether, in vegetable life, there could be any need to excrete, or any advantage gained by throwing off such products; and, while the popular name of catch-fly, given to several common species of *Silene*, indicates long familiarity with the fact, probably no one ever imagined that the swarms of small insects which perish upon these sticky surfaces were ever turned to account by the plant. In many such cases, no doubt they perish as uselessly as when attracted into the flame of a candle. In the tobacco-plant, for instance, Mr. Darwin could find no evidence that the glandular hairs absorb animal matter. But Darwinian philosophy expects all gradations between casualty and complete adaptation. It is most probable that any thin-walled vegetable structure which secretes may also be capable of absorbing under favorable conditions. The myriads of exquisitely-constructed glands of the Chinese primrose are not likely to be functionless. Mr. Darwin ascertained by direct experiment that they promptly absorb carbonate of ammonia, both in watery solution and in vapor. So, since rain-water usually contains a small percentage of ammonia, a use for these glands becomes apparent—one completely congruous with that of absorbing any animal matter, or products of its decomposition, which may come in their way through the occasional entanglement of insects in their viscid secretion. In several saxifrages—not very distant relatives of *Drosera*—the viscid glands equally manifested the power of absorption.

To trace a gradation between a simply absorbing hair with a glutinous tip, through which the plant may perchance derive slight contingent advantage, and the tentacles of a sundew, with their exquisite and associated adaptations, does not much lessen the wonder nor explain the phenomena. After all, as Mr. Darwin modestly concludes, "we see how little has been made out in comparison with what remains unexplained and unknown." But all this must be allowed to be an important contribution to the doctrine of the gradual acquirement of uses and functions, and hardly to find conceivable explanation upon any other hypothesis.

There remains one more mode in which plants of the higher grade are known to prey upon animals; namely, by means of pitchers, urns, or tubes, in which insects and the like are drowned or confined, and either macerated or digested. To this Mr. Darwin barely alludes on the last page of the present volume. The main facts known respecting the American pitcher-plants have, as was natural, been ascertained in this country; and we gave an abstract, two years ago, of our then incipient knowledge. Much has been learned since, although all the observations have been of a desultory character. If space permitted, an instructive narrative might be drawn up, as well of the economy of the *Sarracenias* as of how we came to know what we do of it. But the very little we have room for will be strictly supplementary to our former article.

The pitchers of our familiar Northern *Sarracenia*, which is likewise Southern, are open-mouthed; and, although they certainly secrete some liquid when young, must derive most of the water they ordinarily contain from rain. How insects are attracted is unknown, but the water abounds with their drowned bodies and decomposing remains.

In the more southern *S. flava*, the long and trumpet-shaped pitchers evidently depend upon the liquid which they themselves secrete, although at maturity, when the hood becomes erect, rain may somewhat add to it. This species, as we know, allures insects by a peculiar sweet exudation within the orifice; they fall in and perish, though seldom by drowning, yet few are able to escape; and their decomposing remains accumulate in the narrow bottom of the vessel. Two other long-tubed species of the Southern States are similar in these respects. There is another, *S. psittacina*, the parrot-headed species, remarkable for the cowl-shaped hood so completely inflexed over the mouth of the small pitcher that no rain can possibly enter. Little is known, however, of the efficiency of this species as a fly-catcher; but its conformation has a morphological interest, leading up, as it does, to the Californian type of pitcher presently to be mentioned.

But the remaining species, *S. variolaris*, is the most wonderful of our pitcher-plants in its adaptations for the capture of insects. The inflated and mottled lid or hood overarches the ample orifice of the tubular pitcher sufficiently to ward off the rain, but not to obstruct the free access of flying insects. Flies, ants, and most insects, glide and fall from the treacherous smooth throat into the deep well below, and never escape. They are allured by a sweet secretion just within the orifice—which was discovered and described long ago, and the knowledge of it wellnigh forgotten until recently. And, finally, Dr. Mellichamp, of South Carolina, two years ago made the capital discovery that, during the height of the season, this lure extends from the orifice down nearly to the ground, a length of a foot or two, in the form of a honeyed line or narrow trail on the edge of the wing-like border which is conspicuous in all these species, although only in this one, so far as known, turned to such account. Here, one would say, is a special adaptation to ants and such terrestrial and creeping insects. Well, long before this sweet trail was known, it was remarked by the late Prof. Wyman and others that the pitchers of this species, in the savannahs of Georgia and Florida, contain far more ants than they do of all other insects put together.

Finally, all this is essentially repeated in the peculiar Californian pitcher-plant (*Darlingtonia*), a genus of the same natural family, which captures insects in great variety, enticing them by a sweetish secretion over the whole inside of the inflated hood and that of a curious forked. appendage, resembling a fish-tail, which overhangs the orifice. This orifice is so concealed that it can be seen and approached only from below, as if—the casual observer might infer—to escape visitation. But dead insects of all kinds, and their decomposing remains, crowd the cavity and saturate the liquid therein contained, enticed, it is said, by a peculiar odor, as well as by the sweet lure which is at some stages so abundant as to drip from the tips of the overhanging appendage. The principal observations upon this pitcher-plant in its native habitat have been made by Mrs. Austin, and only some of the earlier ones have thus far been published by Mr. Canby. But we are assured that in this, as in the *Sarracenia variolaris*, the sweet exudation extends at the proper season from the orifice down the wing nearly to the ground, and that ants follow this honeyed pathway to their destruction. Also, that the watery liquid in the pitcher, which must be wholly a secretion, is much increased in quantity after the capture of insects.

It cannot now well be doubted that the animal matter is utilized by the plant in all these cases, although most probably only after maceration or decomposition. In some of them even digestion, or at least the absorption of undecomposed soluble animal juices, may be suspected; but there is no proof of it. But, if pitchers of the Sarracenia family are only macerating vessels, those of *Nepenthes*—the pitchers of the Indian Archipelago, familiar in conservatories—seem to be stomachs. The investigations of the President of the Royal Society, Dr. Hooker, although incomplete, wellnigh demonstrate that these not only allure insects by a sweet secretion at the rim and upon the lid of the cup, but also that their capture, or the presence of other partly soluble animal matter, produces an increase and an acidulation of the contained watery liquid, which thereupon becomes capable of acting in the manner of that of *Drosera* and *Dionæa*, dissolving flesh, albumen, and the like.

After all, there never was just ground for denying to vegetables the use of animal food. The fungi are by far the most numerous family of plants, and they all live upon organic matter, some upon dead and decomposing, some upon living, some upon both; and the number of those that feed upon living animals is large. Whether these carnivorous propensities of higher plants which so excite our wonder be regarded as survivals of ancestral habits, or as comparatively late acquirements, or even as special endowments, in any case what we have now learned of them goes to strengthen the conclusion that the whole organic world is akin.

The volume upon "The Movements and Habits of Climbing Plants" is a revised and enlarged edition of a memoir communicated to the Linnæan Society in 1865, and published in the ninth volume of its Journal. There was an extra impression, but, beyond the circle of naturalists, it can hardly have been much known at first-hand. Even now, when it is made a part of the general Darwinian literature, it is unlikely to be as widely read as the companion volume which we have been reviewing; although it is really a more readable book, and well worthy of far more extended notice at our hands than it can now receive. The reason is obvious. It seems as natural that plants should climb as it does unnatural that any should take animal food. Most people, knowing that some plants "twine with the sun," and others "against the sun," have an idea that the sun in some way causes the twining; indeed, the notion is still fixed in the popular mind that the same species twines in opposite directions north and south of the equator.

Readers of this fascinating treatise will learn, first of all, that the sun has no influence over such movements directly, and that its indirect influence is commonly adverse or disturbing, except the heat, which quickens vegetable as it does animal life. Also, that climbing is accomplished by powers and actions as unlike those generally predicated of the vegetable kingdom as any which have been brought to view in the preceding volume. Climbing plants "feel" as well as "grow and live;" and they also manifest an automatism which is perhaps more wonderful than a response by visible movement to an external irritation. Nor do plants grow up their supports, as is unthinkingly supposed; for, although only growing or newly-grown parts act in climbing, the climbing and the growth are entirely distinct. To this there is one exception—an instructive one, as showing how one action passes into another, and how the same result may be brought about in different ways—that of stems which climb by rootlets, such as of ivy and trumpet-creeper. Here the stem ascends by growth alone, taking upward direction, and is fixed by rootlets as it grows. There is no better way of climbing walls, precipices, and large tree-trunks.

But small stems and similar supports are best ascended by twining; and this calls out powers of another and higher order. The twining stem does not grow around its support, but winds around it, and it does this by a movement the nature of which is best observed in stems which have not yet reached their support, or have overtopped it and stretched out beyond it. Then it may be seen that the extending summit, reaching farther and farther as it grows, is making free circular sweeps, by night as well as by day, and irrespective of external circumstances, except that warmth accelerates the movement, and that the general tendency of young stems to bend toward the light may, in case of lateral illumination, accelerate one-half the circuit while it equally retards the other. The arrest of the revolution where the supporting body is struck, while the portion beyond continues its movement, brings about the twining. As to the proximate cause of this sweeping motion, a few simple experiments prove that it results from the bowing or bending of the free summit of the stem into a more or less horizontal position (this bending being successively to every point of the compass, through an action which circulates around the stem in the direction of the sweep), and of the consequent twining, i. e., "with the sun," or with the movement of the hands of a watch, in the hop, or in the opposite direction in pole-beans and most twiners.

Twining plants, therefore, ascend trees or other stems by an action and a movement of their own, from which they derive advantage. To plants liable to be overshadowed by more robust companions, climbing is an economical method of obtaining a freer exposure to light and air with the smallest possible expenditure of material. But twiners have one disadvantage: to rise ten feet they must produce fifteen feet of stem or thereabouts, according to the diameter of the support, and the openness or closeness of the coil. A rootlet-climber saves much in this respect, but has a restricted range of action, and other disadvantages.

There are two other modes, which combine the utmost economy of material with freer range of action. There are, in the first place, leaf-climbers of various sorts, agreeing only in this, that the duty of laying hold is transferred to the leaves, so that the stem may rise in a direct line. Sometimes the blade or leaflets, or some of them, but more commonly their slender stalks, undertake the work, and the plant rises as a boy ascends a tree, grasping first with one hand or arm, then with the other. Indeed, the comparison, like the leaf-stalk, holds better than would be supposed; for the grasping of the latter is not the result of a blind groping in all directions by a continuous movement, but of a definite sensitiveness which acts only upon the occasion. Most leaves make no regular sweeps; but when the stalks of a leaf-climbing species come into prolonged contact with any fitting extraneous body, they slowly incurve and make a turn around it, and then commonly thicken and harden until they attain a strength which may equal that of the stem itself. Here we have the faculty of movement to a definite end, upon external irritation, of the same nature with that displayed by *Dionæa* and *Drosera*, although slower for the most part than even in the latter. But the movement of the hour-hand of the clock is not different in nature or cause from that of the second-hand.

Finally—distribution of office being, on the whole, most advantageous and economical, and this, in the vegetable kingdom, being led up to by degrees—we reach, through numerous gradations, the highest style of climbing plants in the tendril-climber. A tendril, morphologically, is either a leaf or branch of stem, or a portion of one, specially organized for climbing. Some tendrils simply turn away from light, as do those of grape-vines, thus taking the direction in which some supporting object is likely to be encountered; most are indifferent to light; and many revolve in the manner of the summit of twining stems. As the stems which bear these highly-endowed tendrils in many cases themselves also revolve more or less, though they seldom twine, their reach is the more extensive; and to this endowment of automatic movement most tendrils add the other faculty, that of incurving and coiling upon prolonged touch, or even brief contact, in the highest degree. Some long tendrils, when in their best condition, revolve so rapidly that the sweeping movement may be plainly seen; indeed, we have seen a quarter-circuit in a *Passiflora sicyoides* accomplished in less than a minute, and the half-circuit in ten minutes; but the other half (for a reason alluded to in the next paragraph) takes a much longer time. Then, as to the coiling upon contact, in the case first noticed in this country,[1] in the year 1858, which Mr. Darwin mentions as having led him into this investigation, the tendril of *Sicyos* was seen to coil within half a minute after a stroke with the hand, and to make a full turn or more within the next minute; furnishing ocular evidence that tendrils grasp and coil in virtue of sensitiveness to contact, and, one would suppose, negativing Sachs's recent hypothesis that all these movements are owing "to rapid growth on the side opposite to that which becomes concave"—a view to which Mr. Darwin objects, but not so strongly as he might. The tendril of this sort, on striking some fitting object, quickly curls round and firmly grasps it; then, after some hours, one side shortening or remaining short in proportion to the other, it coils into a spire, dragging the stem up to its support, and enabling the next tendril above to secure a readier hold.

In revolving tendrils perhaps the most wonderful adaptation is that by which they avoid attachment to, or winding themselves upon, the ascending summit of the stem that bears them. This they would inevitably do if they continued their sweep horizontally. But

[1] [A. Gray, in "Proceedings of the American Academy of Arts and Sciences," vol. iv., p. 98; and *American Journal of Science and the Arts*, March, 1859, p. 278.]

when in its course it nears the parent stem the tendril moves slowly, as if to gather strength, then stiffens and rises into an erect position parallel with it, and so passes by the dangerous point; after which it comes rapidly down to the horizontal position, in which it moves until it again approaches and again avoids the impending obstacle.

Climbing plants are distributed throughout almost all the natural orders. In some orders climbing is the rule, in most it is the exception, occurring only in certain genera. The tendency of stems to move in circuits—upon which climbing more commonly depends, and out of which it is conceived to have been educed—is manifested incipiently by many a plant which does not climb. Of those that do there are all degrees, from the feeblest to the most efficient, from those which have no special adaptation to those which have exquisitely-endowed special organs for climbing. The conclusion reached is, that the power "is inherent, though undeveloped, in almost every plant;" "that climbing plants have utilized and perfected a widely-distributed and incipient capacity, which, as far as we can see, is of no service to ordinary plants."

Inherent powers and incipient manifestations, useless to their possessors but useful to their successors—this, doubtless, is according to the order of Nature; but it seems to need something more than natural selection to account for it.

XII.

DURATION AND ORIGINATION OF RACE AND SPECIES.—IMPORT OF SEXUAL REPRODUCTION.

I.

Do Varieties wear out, or tend to wear out?

(New York Tribune, *and* American Journal of Science and the Arts, *February*, 1875.)

This question has been argued from time to time for more than half a century, and is far from being settled yet. Indeed, it is not to be settled either way so easily as is sometimes thought. The result of a prolonged and rather lively discussion of the topic about forty years ago in England, in which Lindley bore a leading part on the negative side, was, if we rightly remember, that the nays had the best of the argument. The deniers could fairly well explain away the facts adduced by the other side, and evade the force of the reasons then assigned to prove that varieties were bound to die out in the course of time. But if the case were fully re-argued now, it is by no means certain that the nays would win it. The most they could expect would be the Scotch verdict, "not proven." And this not because much, if any, additional evidence of the actual wearing out of any vari ety has turned up since, but because a presumption has been raised under which the evidence would take a bias the other way. There is now in the minds of scientific men some reason to expect that certain varieties would die out in the long run, and this might have an important influence upon the interpretation of the facts. Curiously enough, however, the recent discussions to which our attention has been called seem, on both sides, to have overlooked this.

But, first of all, the question needs to be more specifically stated. There are varieties and varieties. They may, some of them, disappear or deteriorate, but yet not wear out—not come to an end from any inherent cause. One might even say, the younger they are the less the chance of survival unless well cared for. They may be smothered out by the adverse force of superior numbers; they are even more likely to be bred out of existence by unprevented cross-fertilization, or to disappear from mere change of fashion. The question, however, is not so much about reversion to an ancestral state, or the falling off of a high-bred stock into an inferior condition. Of such cases it is enough to say that, when a variety or strain, of animal or vegetable, is led up to unusual fecundity or of size or product of any organ, for our good, and not for the good of the plant or animal itself, it can be kept so only by high feeding and exceptional care; and that with high feeding and artificial appliances comes vastly increased liability to disease, which may practically annihilate the race. But then the race, like the bursted boiler, could not be said to wear out, while if left to ordinary conditions, and allowed to degenerate back into a more natural if less useful state, its hold on life would evidently be increased rather than diminished.

As to natural varieties or races under normal conditions, sexually propagated, it could readily be shown that they are neither more nor less likely to disappear from any inherent cause than the species from which they originated. Whether species wear out, i. e., have their rise, culmination, and decline, from any inherent cause, is wholly a geological and very speculative problem, upon which, indeed, only vague conjectures can be offered. The matter actually under discussion concerns cultivated domesticated varieties only, and, as to plants, is covered by two questions.

First, *Will races propagated by seed*, being so fixed that they come true to seed, and purely bred (not crossed with any other sort), continue so indefinitely, or *will they run out in time*—not die out, perhaps, but lose their distinguishing characters? Upon this, all we are able to say is that we know no reason why they should wear out or deteriorate from any inherent cause. The transient existence or the deterioration and disappearance of many such races are sufficiently accounted for otherwise; as in the case of extraordinarily exuberant varieties, such as mammoth fruits or roots, by increased liability to disease, already adverted to, or by the failure of the high feeding they demand. A common cause, in ordinary cases, is cross-breeding, through the agency of wind or insects, which is difficult to guard against. Or they go out of fashion and are superseded by others thought to be better, and so the old ones disappear.

Or, finally, they may revert to an ancestral form.

As offspring tend to resemble grandparents almost as much as parents, and as a line of close-bred ancestry is generally prepotent, so newly-originated varieties have always a tendency to reversion. This is pretty sure to show itself in some of the progeny of the earlier generations, and the breeder has to guard against it by rigid selection. But the older the variety is—that is, the longer the series of generations in which it has come true from seed—the less the chance of reversion: for now, to be like the immediate parents, is also to be like a long line of ancestry; and so all the influences concerned—that is, both parental and ancestral heritability—act in one and the same direction. So, since the older a race is the more reason it has to continue true, the presumption of the unlimited permanence of old races is very strong.

Of course the race itself may give off new varieties; but that is no interference with the vitality of the original stock. If some of the new varieties supplant the old, that will not be because the unvaried stock is worn out or decrepit with age, but because in wild Nature the newer forms are better adapted to the surroundings, or, under man's care, better adapted to his wants or fancies.

The second question, and one upon which the discussion about the wearing out of varieties generally turns, is, *Will varieties propagated from buds, i. e., by division, grafts, bulbs, tubers, and the like, necessarily deteriorate and die out?* First, Do they die out as a matter of fact? Upon this, the testimony has all along been conflicting. Andrew Knight was sure that they do, and there could hardly be a more trustworthy witness.

"The fact," he says, fifty years ago, "that certain varieties of some species of fruit which have been long cultivated cannot now be made to grow in the same soils and under the same mode of management, which was a century ago so perfectly successful, is placed beyond the reach of controversy. Every experiment which seemed to afford the slightest prospect of success was tried by myself and others to propagate the old varieties of the apple and pear which formerly constituted the orchards of Herefordshire, without a single healthy or efficient tree having been obtained; and I believe all attempts to propagate these varieties have, during some years, wholly ceased to be made."

To this it was replied, in that and the next generation, that cultivated vines have been transmitted by perpetual division from the time of the Romans, and that several of the sorts, still prized and prolific, are well identified, among them the ancient Græcula, considered to be the modern Corinth or currant grape, which has immemorially been seedless; that the old nonpareil apple was known in the time of Queen Elizabeth; that the white beurré pears of France have been propagated from the earliest times; and that golden pippins, St. Michael pears, and others said to have run out, were still to be had in good condition.

Coming down to the present year, a glance through the proceedings of pomological societies, and the debates of farmers' clubs, brings out the same difference of opinion. The testimony is nearly equally divided. Perhaps the larger number speak of the deterioration and failure of particular old sorts; but when the question turns on "wearing out," the positive evidence of vigorous trees and sound fruits is most telling. A little positive testimony outweighs a good deal of negative. This cannot readily be explained away, while the failures may be, by exhaustion of soil, incoming of disease, or alteration of climate or circumstances. On the other hand, it may be urged that, if a variety of this sort is fated to become decrepit and die out, it is not bound to die out all at once, and everywhere at the same time. It would be expected first to give way wherever it is weakest, from whatever cause. This consideration has an important bearing upon the final question, Are old varieties of this kind on the way to die out on account of their age or any inherent limit of vitality?

Here, again, Mr. Knight took an extreme view. In his essay in the "Philosophical Transactions," published in the year 1810, he propounded the theory, not merely of a natural limit to varieties from grafts and cuttings, but even that they would not survive the natural term of the life of the seedling trees from which they were originally taken. Whatever may have been his view of the natural term of the life of a tree, and of a cutting being merely a part of the individual that produced it, there is no doubt that he laid himself open to the effective replies which were made from all sides at the time, and have lost none of their force since. Weeping-willows, bread-fruits, bananas, sugar-cane, tiger-lilies, Jerusalem artichokes, and the like, have been propagated for a long while in this way, without evident decadence.

Moreover, the analogy upon which his hypothesis is founded will not hold. Whether or not one adopts the present writer's conception, that individuality is not actually reached or maintained in the vegetable world, it is clear enough that a common plant or tree is not an individual in the sense that a horse or man, or any one of the higher animals, is—that it is an individual only in the sense that a branching zoöphyte or mass of coral is. *Solvitur crescendo:* the tree and the branch equally demonstrate that they are not individuals, by being divided with impunity and advantage, with no loss of life, but much increase. It looks odd enough to see a writer like Mr. Sisley reproducing the old hypothesis in so bare a form as this: "I am prepared to maintain that varieties are individuals, and that as they are born they must die, like other individuals. . . . We know that oaks, Sequoias, and other trees, live several centuries, but how many we do not exactly know. But that they must die, no one in his senses will dispute." Now, what people in their senses do dispute is, not that the tree will die, but that other trees, established from its cuttings, will die with it.

But does it follow from this that non-sexually-propagated varieties are endowed with the same power of unlimited duration that is possessed by varieties and species propagated sexually—i. e., by seed? Those who think so jump too soon at their conclusion. For, as to the facts, it is not enough to point out the diseases or the trouble in the soil or the atmosphere to which certain old fruits are succumbing, nor to prove that a parasitic fungus (*Peronospora infestans*) is what is the matter with potatoes. For how else would constitutional debility, if such there be, more naturally manifest itself than in such increased liability or diminished resistance to such attacks? And if you say that, anyhow, such varieties do not die of old age

—meaning that each individual attacked does not die of old age, but of manifest disease—it may be asked in return, what individual man ever dies of old age in any other sense than of a similar inability to resist invasions which in earlier years would have produced no noticeable effect? Aged people die of a slight cold or a slight accident, but the inevitable weakness that attends old age is what makes these slight attacks fatal.

Finally, there is a philosophical argument which tells strongly for some limitation of the duration of non-sexually-propagated forms, one that probably Knight never thought of, but which we should not have expected recent writers to overlook. When Mr. Darwin announced the principle that cross-fertilization between the individuals of a species is the plan of Nature, and is practically so universal that it fairly sustains his inference that no hermaphrodite species continually self-fertilized would continue to exist, he made it clear to all who apprehend and receive the principle that a series of plants propagated by buds only must have weaker hold of life than a series reproduced by seed. For the former is the closest possible kind of close breeding. Upon this ground such varieties may be expected ultimately to die out; but "the mills of the gods grind so exceeding slow" that we cannot say that any particular grist has been actually ground out under human observation.

If it be asked how the asserted principle is proved or made probable, we can here merely say that the proof is wholly inferential. But the inference is drawn from such a vast array of facts that it is well-nigh irresistible. It is the legitimate explanation of those arrangements in Nature to secure cross-fertilization in the species, either constantly or occasionally, which are so general, so varied and diverse, and, we may add, so exquisite and wonderful, that, once propounded, we see that it must be true.[1] What else, indeed, is the meaning and use of sexual reproduction? Not simply increase of numbers; for that is otherwise effectually provided for by budding propagation in plants and many of the lower animals. There are plants, indeed, of the lower sort (such as diatoms), in which the whole multiplication takes place in this way, and with great rapidity. These also have sexual reproduction; but in it two old individuals are always destroyed to make a single new one! Here propagation diminishes the number of individuals fifty per cent. Who can suppose that such a costly process as this, and that all the exquisite arrangements for cross-fertilization in hermaphrodite plants, do not subserve some most important purpose? How and why the union of two organisms, or generally of two very mi-

[1] Here an article would be in place, explaining the arrangements in Nature for cross-fertilization, or wide-breeding, in plants, through the agency, sometimes of the winds, but more commonly of insects; the more so, since the development of the principle, the appreciation of its importance, and its confirmation by abundant facts, are mainly due to Mr. Darwin. But our reviews and notices of his early work "On the Contrivances in Nature for the Fertilization of Orchids by Means of Insects," in 1862, and his various subsequent papers upon other parts of this subject, are either to otechnical or too fragmentary or special to be here reproduced. Indeed, a popular essay is now hardly needed, since the topic has been fully presented, of late years, in the current popular and scientific journals, and in common educational works and text-books, so that it is in the way of becoming a part—and a most inviting part—of ordinary botanical instruction.

nute portions of them, should reënforce vitality, we do not know, and can hardly conjecture. But this must be the meaning of sexual reproduction.

The conclusion of the matter, from the scientific point of view, is, that sexually-propagated varieties or races, although liable to disappear through change, need not be expected to wear out, and there is no proof that they do; but, that non-sexually propagated varieties, though not especially liable to change, may theoretically be expected to wear out, but to be a very long time about it.

II.

Do Species wear out? and if not, why not?

The question we have just been considering was merely whether races are, or may be, as enduring as species. As to the inherently unlimited existence of species themselves, or the contrary, this, as we have said, is a geological and very speculative problem. Not a few geologists and naturalists, however, have concluded, or taken for granted, that species have a natural term of existence—that they culminate, decline, and disappear through exhaustion of specific vitality, or some equivalent internal cause. As might be expected from the nature of the inquiry, the facts which bear upon the question are far from decisive. If the fact that species in general have not been interminable, but that one after another in long succession has become extinct, would seem to warrant this conclusion, the persistence through immense periods of no inconsiderable number of the lower forms of vegetable and animal life, and of a few of the higher plants from the Tertiary period to the present, tells even more directly for the limitless existence of species. The disappearance is quite compatible with the latter view; while the persistence of any species is hardly explicable upon any other. So that, even under the common belief of the entire stability and essential inflexibility of species, extinction is more likely to have been accidental than predetermined, and the doctrine of inherent limitation is unsupported by positive evidence.

On the other hand, it is an implication of the Darwinian doctrine that species are essentially unlimited in existence. When they die out—as sooner or later any species may—the verdict must be accidental death, under stress of adverse circumstances, not exhaustion of vitality; and, commonly, when the species seems to die out, it will rather have suffered change. For the stock of vitality which enables it to vary and survive in changed forms under changed circumstances must be deemed sufficient for a continued unchanged existence under unaltered conditions. And, indeed, the advancement from simpler to more complex, which upon the theory must have attended the diversification, would warrant or require the supposition of increase instead of diminution of power from age to age.

The only case we call to mind which, under the Darwinian view, might be interpreted as a dying out from inherent causes, is that of a species which refuses to vary, and thus lacks the capacity of adaptation to altering conditions. Under altering conditions, this lack would be fatal. But this would be the fatality of *some* species or form in particular, not of species or forms generally, which, for the most part, may and do vary sufficiently, and in varying survive, seemingly none the worse, but rather the better, for their long tenure of life.

The opposite idea, however, is maintained by M. Naudin,[1] in a detailed exposition of his own views of evolution, which differ widely from those of Darwin in most respects, and notably in excluding that which, in our day, gives to the subject its first claim to scientific (as distinguished from purely speculative) attention; namely, natural selection. Instead of the causes or operations collectively personified under this term, and which are capable of exact or probable appreciation, M. Naudin invokes "the two principles of rhythm and of the decrease of forces in Nature." He is a thorough evolutionist, starting from essentially the same point with Darwin; for he conceives of all the forms or species of animals and plants "comme tiré tout entier d'un protoplasma primordial, uniform, instable, éminemment plastique." Also in "l'intégration croissante de la force évolutive à mesure qu'elle se partage dans les formes produites, et la décroissance proportionelle de la plasticité de ces formes à mesure qu'elles s'éloignent davantage de leur origine, et qu'elles sont mieux arrêtées." As they get older, they gain in fixity through the operation of the

[1] "Les Espèces affines et la Théorie de l'Évolution," par Charles Naudin, Membre de l'Institut, in *Bulletin de la Société Botanique de France*, tome xxi., pp. 240–272, 1874. *See* also *Comptes Rendus*, September 27 and October 4, 1875, reproduced in "Annales des Sciences Naturelles," 1876, pp. 73–81.

fundamental law of inheritance; but the species, like the individual, loses plasticity and vital force. To continue in the language of the original:

"C'est dire qu'il y a eu, pour l'ensemble du monde organique, une période de formation où tout était changeant et mobile, une phase analogue à la vie embryonnaire et à la jeunesse de chaque être particulier; et qu'à cet âge de mobilité et de croissance a succédé une période de stabilité, au moins relative, une sorte d'âge adulte, où la force évolutive, ayant achevé son œuvre, n'est plus occupée qu'à la maintenir, sans pouvoir produire d'organismes nouveaux. Limitée en quantité, comme toutes les forces en jeu dans une planète ou dans un système sidéral tout entier, cette force n'a pu accomplir qu'un travail limité; et du même qu'un organisme, animal ou végétal, ne croit pas indéfiniment et qu'il s'arrête à des proportions que rien ne peut faire dépasser, de même aussi l'organisme total de la nature s'est arrêté à un état d'équilibre, dont la durée, selon toutes vraisemblances, doit être beaucoup plus longue que celle de la phase de développement et de croissance.

A fixed amount of "evolutive force" is given, to begin with. At first enormous, because none has been used up in work, it is necessarily enfeebled in the currents into which the stream divides, and the narrower and narrower channels in which it flows with slowly-diminishing power. Hence the limited although very unequal duration of all individuals, of all species, and of all types of organization. A multitude of forms have disappeared already, and the number of species, far from increasing, as some have believed, must, on the contrary, be diminishing. Some species, no doubt, have suffered death by violence or accident, by geological changes, local alteration of the conditions, or the direct or indirect attacks of other species; but these have only anticipated their fate, for M. Naudin contends that most of the extinct species have died a natural death from exhaustion of force, and that all the survivors are on the way to it. The great timepiece of Nature was wound up at the beginning, and is running down. In the earlier stages of great plasticity and exuberant power, diversification took place freely, but only in definite lines, and species and types multiplied. As the power of survival is inherently limited, still more the power of change: this diminishes in time, if we rightly apprehend the idea, partly through the waning of vital force, partly through the fixity acquired by heredity —like producing like, the more certainly in proportion to the length and continuity of the ancestral chain. And so the small variations of species which we behold are the feeble remnants of the pristine plasticity and an exhausted force.[1] This force of variation or origination of forms has acted rhythmically or intermittently, because each movement was the result of the rupture of an equilibrium, the liber-

[1] In noticing M. Naudin's paper in the *Comptes Rendus*, now reprinted in the "Annales des Sciences Naturelles," entitled "Variation désordonnée des Plantes Hybrides et Deductions qu'on peut en tirer," we were at a loss to conceive why he attributed all present variation of species to atavism, i. e., to the reappearance of ancestral characters (*American Journal of Science*, February, 1876). His anterior paper was not then known to us; from which it now appears that this view comes in as a part of the hypothesis of extreme plasticity and variability at the first, subsiding at length into entire fixity and persistence of character. According to which, it is assumed that the species of our time have lost all power of original variation, but can still reproduce some old ones—some reminiscences, as it were, of youthful vagaries—in the way of atavism.

ation of a force which till then was retained in a potential state by some opposing force or obstacle, overcoming which, it passes to a new equilibrium, and so on. Hence alternations of dynamic activity and static repose, of origination of species and types, alternated with periods of stability or fixity. The timepiece does not run down regularly, but "la force procède par saccades; et . . . . par pulsations d'autant plus énergiques que la nature était plus près de son commencement."

Such is the hypothesis. For a theory of evolution, this is singularly unlike Darwin's in most respects, and particularly in the kind of causes invoked and speculations indulged in. But we are not here to comment upon it beyond the particular point under consideration, namely, its doctrine of the inherently limited duration of species. This comes, it will be noticed, as a deduction from the modern physical doctrine of the equivalence of force. The reasoning is ingenious, but, if we mistake not, fallacious.

To call that "evolutive force" which produces the change of one kind of plant or animal into another, is simple and easy, but of little help by way of explanation. To homologize it with physical force, as M. Naudin's argument requires, is indeed a step, and a hardy one; but it quite invalidates the argument. For, if the "evolutive force" is a part of the physical force of the universe, of which, as he reminds us, the sum is fixed and the tendency is toward a stable equilibrium in which all change is to end, then this evolutive was derived from the physical force; and why not still derivable from it? What is to prevent its replenishment in vegetation, *pari passu* with that great operation in which physical force is stored up in vegetable organisms, and by the expenditure or transformation of which their work, and that of all animals, is carried on? Whatever be the cause (if any there be) which determines the decadence and death of species, one cannot well believe that it is a consequence of a diminution of their proper force by plant-development and division; for instance, that the sum of what is called vital force in a full-grown tree is not greater, instead of less, than that in the seedling, and in the grove greater than in the single parental tree. This power, if it be properly a force, is doubtless as truly derived from the sunbeam as is the power which the plant and animal expend in work. Here, then, is a source of replenishment as lasting as the sun itself, and a ground—so far as a supply of force is concerned—for indefinite duration. For all that any one can mean by the indefinite existence of species is, that they may (for all that yet appears) continue while the external conditions of their being or well-being continue.

Perhaps, however, M. Naudin does not mean that "evolutive force," or the force of vitality, is really homologous with common physical force, but only something which may be likened to it. In that case the parallel has only a metaphorical value, and the reason why variation must cease and species die out is still to seek. In short, if that which continues the series of individuals in propagation, whether like or unlike the parents, be a force in the physical sense of the term, then there is abundant provision in Nature for its indefinite replenishment. If, rather, it be a part or phase of that something which directs and determines the expenditure of force, then it is not subject to the laws of the latter, and there is no ground for inferring its exhaustibility. The limited vitality is an unproved and unprovable conjecture. The evolutive force, dying out in the using, is either the same conjecture repeated, or a misapplied analogy.

After all—apart from speculative analogies—the only evidences we possess which indicate a tendency in species to die out, are those to which Mr. Darwin has called attention. These are, first, the observed deterioration which results, at least in animals, from continued breeding in and in, which may possibly be resolvable into cumulative heritable disease; and, secondly, as already stated (p. 346), what may be termed the sedulous and elaborate pains everywhere taken in Nature to prevent close breeding—arrangements which are particularly prominent in plants, the greater number of which bear hermaphrodite blossoms. The importance of this may be inferred from the universality, variety, and practical perfection of the arrangements which secure the end; and the inference may fairly be drawn that this is the physiological import of sexes.

It follows from this that there is a tendency, seemingly inherent, in species as in individuals, to die out; but that this tendency is counteracted or checked by sexual wider breeding, which is, on the whole, amply secured in Nature, and which in some way or other reënforces vitality to such an extent as to warrant Darwin's inference that "some unknown great good is derived from the union of individuals which have been kept distinct for many generations." Whether this reënforcement is a complete preventive of decrepitude in species, or only a palliative, is more than we can determine. If the latter, then existing species and their derivatives must perish in time, and the earth may be growing poorer in species, as M. Naudin supposes, through mere senility. If the former, then the earth, if not even growing richer, may be expected to hold its own, and extant species or their derivatives should last as long as the physical world lasts and affords favorable conditions. General analogies seem to favor the former view. Such facts as we possess, and the Darwinian hypothesis, favor the latter.

XIII.

EVOLUTIONARY TELEOLOGY.

When Cuvier spoke of the "combination of organs in such order that they may be in consistence with the part which the animal has to play in Nature," his opponent, Geoffroy St.-Hilaire, rejoined, "I know nothing of animals which have to play a part in Nature." The discussion was a notable one in its day. From that time to this, the reaction of morphology against "final causes" has not rarely gone to the extent of denying the need and the propriety of assuming ends in the study of animal and vegetable organizations. Especially in our own day, when it became apparent that the actual use of an organ might not be the fundamental reason of its existence—that one and the same organ, morphologically considered, was modified in different cases to the most diverse uses, while intrinsically different organs subserved identical functions, and consequently that use was a fallacious and homology the surer guide to correct classification—it was not surprising that teleological ideas nearly disappeared from natural history. Probably it is still generally thought that the school of Cuvier and that of St.-Hilaire have neither common ground nor capability of reconcilement.

In a review of Darwin's volume on the "Fertilization of Orchids"[1] (too technical and too detailed for reproduction here), and later in a brief sketch of the character of his scientific work (art. x., p. 284), we expressed our sense of the great gain to science from his having brought back teleology to natural history. In Darwinism, usefulness and purpose come to the front again as working principles of the first order; upon them, indeed, the whole system rests.

To most, this restoration of teleology has come from an unexpected quarter, and in an unwonted guise; so that the first look of it is by no means reassuring to the minds of those who cherish theistic views of Nature. Adaptations irresistibly suggesting purpose had their supreme application in natural theology. Being manifold, particular, and exquisite, and evidently inwrought into the whole system of the organic world, they were held to furnish irrefragable as well as independent proof of a personal designer, a divine originator of Nature. By a confusion of thought, now obvious, but at the time not unnatural, they were also regarded as proof of a direct execution of the contriver's purpose in the creation of each organ and organism, as it were, in the manner man contrives and puts together a machine—an idea which has been set up as the orthodox doctrine, but which to St. Augustine and other learned Christian fathers would have savored of heterodoxy.

In the doctrine of the origination of species through natural selection, these adaptations appear as the outcome rather than as the motive, as final results rather

[1] London, 1862.

than final causes. Adaptation to use, although the very essence of Darwinism, is not a fixed and inflexible adaptation, realized once for all at the outset; it includes a long progression and succession of modifications, adjusting themselves to changing circumstances, under which they may be more and more diversified, specialized, and in a just sense perfected. Now, the question is, Does this involve the destruction or only the reconstruction of our consecrated ideas of teleology? Is it compatible with our seemingly inborn conception of Nature as an ordered system? Furthermore, and above all, can the Darwinian theory itself dispense with the idea of purpose, in the ordinary sense of the word, as tantamount to design?

From two opposing sides we hear the first two questions answered in the negative. And an affirmative response to the third is directly implied in the following citation:

"The word *purpose* has been used in a sense to which it is, perhaps, worth while to call attention. Adaptation of means to an end may be provided in two ways that we at present know of: by processes of natural selection, and by the agency of an intelligence in which an image or idea of the end preceded the use of the means. In both cases the existence of the adaptation is accounted for by the necessity or utility of the end. It seems to me convenient to use the word purpose as meaning generally the end to which certain means are adapted, both in these two cases and in any other that may hereafter become known, provided only that the adaptation is accounted for by the necessity or utility of the end. And there seems no objection to the use of the phrase 'final cause' in this wider sense, if it is to be kept at all. The word 'design' might then be kept for the special case of adaptation by an intelligence. And we may then say that, since the process of natural selection has been understood, *purpose* has ceased to suggest design to instructed people, except in cases where the agency of man is independently probable."—P. C. W., in the *Contemporary Review* for September, 1875, p. 657.

The distinction made by this anonymous writer is convenient and useful, and his statement clear. We propose to adopt this use of the terms *purpose* and *design*, and to examine the allegation. The latter comes to this: "Processes of natural selection" exclude "the agency of an intelligence in which the image or idea of the end precedes the use of the means;" and since the former have been understood "purpose has ceased to suggest design to instructed people, except in cases where the agency of man is independently probable." The maxim "*L'homme propose, Dieu dispose*," under this reading means that the former has the monopoly of design, while the latter accomplishes without designing. Man's works alone suggest design.

But it is clear to us that this monopoly is shared with certain beings of inferior grade. Granting that quite possibly the capture of flies for food by *Dionæa* and the sundews may be attributed to purpose apart from design (if it be practicable in the last resort to maintain this now convenient distinction), still their capture by a spider's-web, and by a swallow on the wing, can hardly "cease to suggest design to instructed people." And surely, in coming at his master's call, the dog fulfills his own design as well as that of his master; and so of other actions and constructions of brute animals.

Without doubt so acute a writer has a clear and

sensible meaning; so we conclude that he regards brutes as automata, and was thinking of design as co-extensive merely with general conceptions. Not concerning ourselves with the difficulty he may have in drawing a line between the simpler judgments and affections of man and those of the highest-endowed brutes, we subserve our immediate ends by remarking that the automatic theory would seem to be one which can least of all dispense with design, since, either in the literal or current sense of the word, undesigned automatism is, as near as may be, a contradiction in terms. As the automaton man constructs manifests the designs of its maker and mover, so the more efficient automata which man did not construct would not legitimately suggest less than human intelligence. And so all adaptations in the animal and vegetable world which irresistibly suggest purpose (in the sense now accepted) would also suggest design, and, under the law of parsimony, claim to be thus interpreted, unless some other hypothesis will better account for the facts. We will consider, presently, if any other does so.

We here claim only that some beings other than men design, and that the adaptations of means to ends in the structure of animals and plants, in so far as they carry the marks of purpose, carry also the implication of having been designed. Also, that the idea or hypothesis of a designing mind, as the author of Nature—however we came by it—having possession of the field, and being one which man, himself a designer, seemingly must needs form, cannot be rivaled except by some other equally adequate for explanation, or displaced except by showing the illegitimacy of the inference. As to the latter, is the common apprehension and sense of mankind in this regard well grounded? Can we rightly reason from our own intelligence and powers to a higher or a supreme intelligence ordering and shaping the system of Nature?

A very able and ingenious writer upon "The Evidences of Design in Nature," in the *Westminster Review* for July, 1875, maintains the negative. His article may be taken as the argument in support of the position assumed by "P. C. W.," in the *Contemporary Review* above cited. It opens with the admission that the orthodox view is the most simple and apparently convincing, has had for centuries the unhesitating assent of an immense majority of thinkers, and that the latest master-writer upon the subject disposed to reject it, namely, Mill, comes to the conclusion that, "in the present state of our knowledge, the adaptations in Nature afford a large balance of probability in favor of creation by intelligence." It proceeds to attack not so much the evidence in favor of design as the foundation upon which the whole doctrine rests, and closes with the prediction that sooner or later the superstructure must fall. And, truly, if his reasonings are legitimate, and his conclusions just, "Science has laid the axe to the tree."

"Given a set of marks which we look upon in human productions as unfailing indications of design," he asks, "is not the inference equally legitimate when we recognize these marks in Nature? To gaze on such a universe as this, to feel our hearts exult within us in the fullness of existence, and to offer in explanation of such beneficent provision no other word but

Chance, seems as unthankful and iniquitous as it seems absurd. Chance produces nothing in the human sphere; nothing, at least, that can be relied upon for good. Design alone engenders harmony, consistency; and Chance not only never is the parent, but is constantly the enemy of these. How, then, can we suppose Chance to be the author of a system in which everything is as regular as clock-work? . . . . The hypothesis of Chance is inadmissible."

There is, then, in Nature, an order; and, in "P. C. W.'s" sense of the word, a manifest purpose. Some sort of conception as to the cause of it is inevitable, that of design first and foremost. "Why"—the *Westminster Reviewer* repeats the question—"why, if the marks of utility and adaptation are conclusive in the works of man, should they not be considered equally conclusive in the works of Nature?" His answer appears to us more ingenious than sound. Because, referring to Paley's watch,—

"The watch-finder is not guided solely in his inference by marks of adaptation and utility; he would recognize design in half a watch, in a mere fragment of a watch, just as surely as in a whole time-keeper. . . . Two cog-wheels, grasping each other, will be thought conclusive evidence of design, quite independently of any use attaching to them. And the inference, indeed, is perfectly correct; only it is an inference, not from a mark of design, properly so called, but from a mark of human workmanship. . . . No more is needed for the watch-finder, since all the works of man are, at the same time, products of design; but a great deal more is requisite for us, who are called upon by Paley to recognize design in works in which this stamp, this label of human workmanship, is wanting. The mental operation required in the one case is radically different from that performed in the other; there is no parallel, and Paley's demonstration is totally irrelevant."[1]

[1] Hume, in his "Essays," anticipated this argument. But he did

But, surely, all human doings are not "products of design;" many are contingent or accidental. And why not suppose that the finder of the watch, or of the watch-wheel, infers *both* design and human workmanship? The two are mutually exclusive only on the supposition that man alone is a designer, which is simply begging the question in discussion. If the watch-finder's attention had been arrested by

not rest on it. His matured convictions appear to be expressed in statements such as the following, here cited at second hand from Jackson's "Philosophy of Natural Theology," a volume to which a friend has just called our attention:

"Though the stupidity of men," writes Hume, "barbarous and uninstructed, be so great that they may not see a sovereign author in the more obvious works of Nature, to which they are so much familiarized, yet it scarce seems possible that any one of good understanding should reject that idea, when once it is suggested to him. A purpose, an intention, a design, is evident in everything; and when our comprehension is so far enlarged as to contemplate the first rise of this visible system, we must adopt, with the strongest conviction, the idea of some intelligent cause or author. The uniform maxims, too, which prevail throughout the whole frame of the universe, naturally, if not necessarily, lead us to conceive this intelligence as single and undivided, where the prejudices of education oppose not so reasonable a theory. Even the contrarieties of Nature, by discovering themselves everywhere, become proofs of some consistent plan, and establish one single purpose or intention, however inexplicable and incomprehensible."—("Natural History of Religion," xv.)

"In many views of the universe, and of its parts, particularly the latter, the beauty and fitness of final causes strike us with such irresistible force that all objections appear (what I believe they really are) mere cavils and sophisms."—("Dialogues concerning Natural Religion," Part X.)

"The order and arrangement of Nature, the curious adjustment of final causes, the plain use and intention of every part and organ, all these bespeak in the clearest language an intelligent cause or author."—(Ibid., Part IV.)

a different object, such as a spider's web, he would have inferred both design and non-human workmanship. Of some objects he might be uncertain whether they were of human origin or not, without ever doubting they were designed, while of others this might remain doubtful. Nor is man's recognition of human workmanship, or of any other, dependent upon his comprehending how it was done, or what particular ends it subserves. Such considerations make it clear that "the label of human workmanship" is not the generic stamp from which man infers design. It seems equally clear that "the mental operation required in the one case" is not so radically or materially "different from that performed in the other" as this writer would have us suppose. The judgment respecting a spider's web, or a trap-door spider's dwelling, would be the very same in this regard if it preceded, as it occasionally might, all knowledge of whether the object met with were of human or animal origin. A dam across a stream, and the appearance of the stumps of trees which entered into its formation, would suggest design quite irrespective of and antecedent to the considerable knowledge or experience which would enable the beholder to decide whether this was the work of men or of beavers. Why, then, should the judgment that any particular structure is a designed work be thought illegitimate when attributed to a higher instead of a lower intelligence than that of man? It might, indeed, be so if the supposed observer had no conception of a power and intelligence superior to his own. But it would then be more than "irrelevant;" it would be impossible, except on the supposition that the phenomena would of themselves give rise to such an inference. That it is now possible to make the inference, and, indeed, hardly possible not to make it, is sufficient warrant of its relevancy.

It may, of course, be rejoined that, if this important factor is given, the inference yields no independent argument of a divine creator; and it may also be reasonably urged that the difference between things that are made under our observation and comprehension, and things that grow, but have originated beyond our comprehension, is too wide for a sure inference from the one to the other. But the present question. involves neither of these. It is simply whether the argument for design from adaptations in Nature is relevant, not whether it is independent or sure. It is conceded that the argument is analogical, and the parallel incomplete. But the *gist* is in the points that are parallel or similar. Pulleys, valves, and such-like elaborate mechanical adaptations, cannot differ greatly in meaning, wherever met with.

The opposing argument is repeated and pressed in another form:

"The evidence of design afforded by the marks of adaptation in works of human competence is null and void in the case of creation itself. . . . Nature is full of adaptations; but these are valueless to us as traces of design, unless we know something of the rival adaptations among which an intelligent being might have chosen. To assert that in Nature no such rival adaptations existed, and that in every case the useful function in question could be established by no other instrument but one, is simply to reason in a circle, since it is solely from what we find existing that our notions of possibility and impossibility are drawn. . . . We cannot imagine ourselves in the position of the Creator before his work began, nor examine the materials among which he had to choose, nor count the laws which limited his operations. Here all is dark, and the inference we draw from the seeming perfections of the existing instruments or means is a measure of nothing but our ignorance."

But the question is not about the perfection of these adaptations, or whether others might have been instituted in their place. It is simply whether observed adaptations of intricate sorts, admirably subserving uses, do or do not legitimately suggest to one designing mind that they are the product of some other. If so, no amount of ignorance, or even inconceivability, of the conditions and mode of production could affect the validity of the inference, nor could it be affected by any misunderstanding on our part as to what the particular use or function was; a statement which would have been deemed superfluous, except for the following:

"There is not an organ in our bodies but what has passed, and is still passing, through a series of different and often contradictory interpretations. Our lungs, for instance, were anciently conceived to be a kind of cooling apparatus, a refrigerator; at the close of the last century they were supposed to be a centre of combustion; and nowadays both these theories have been abandoned for a third. . . . Have these changes modified in the slightest degree the supposed evidence of design?"

We have not the least idea why they should. So, also, of complicated processes, such as human digestion, being replaced by other and simpler ones in lower animals, or even in certain plants. If "we argue the necessity of every adaptation solely from the fact that it exists," and that "we cannot mutilate it grossly without injury to the function," we do not "announce triumphantly that digestion is impossible in any way but this," etc., but see equal wisdom and no impugnment of design in any number of simpler adaptations accomplishing equivalent purposes in lower animals.

Finally, adaptation and utility being the only marks of design in Nature which we possess, and adaptation only as subservient to usefulness, the *Westminster Reviewer* shows us how—

"The argument from utility may be equally refuted another way. We found in our discussion of the mark of adaptation that the positive evidence of design afforded by the mechanisms of the human frame was never accompanied by the possibility of negative evidence. We regarded this as a suspicious circumstance, just as the fox, invited to attend the lion in his den, was deterred from his visit by observing that all the foot-tracks lay in one direction. The same suspicious circumstance warns us now. If positive evidence of design be afforded by the presence of a faculty, negative evidence of design ought to be afforded by the absence of a faculty. This, however, is not the case." [Then follows the account of a butterfly, which, from the wonderful power of the males to find the females at a great distance, is conceived to possess a sixth sense.] "Do we consider the deficiency of this sixth sense in man as the slightest evidence against design? Should we be less apt to infer creative wisdom if we had only four senses instead of five, or three instead of four? No, the case would stand precisely as it does now. We value our senses simply because we have them, and because our conception of life as we desire it is drawn from them. But to reason from such value to the origin of our endowment, to argue that our senses must have been given to us by a deity because we prize them, is evidently to move round and round in a vicious circle.

"The same rejoinder is easily applicable to the argument from beauty, which indeed is only a particular aspect of the argument from utility. It is certainly improbable that a random daubing of colors on a canvas will produce a tolerable painting, even should the experiment be continued for thousands of years. Our conception of beauty being given, it is utterly improbable that chance should select, out of the infinity of combinations which form and color may afford, the precise combination which that conception will approve. But the universe is not posterior to our sense of beauty, but antecedent to it: our sense of beauty grows out of what we see; and hence the conformance of our world to our æsthetical conceptions is evidence, not of the world's origin, but of our own."

We are accustomed to hear design doubted on account of certain failures of provision, waste of resources, or functionless condition of organs; but it is refreshingly new to have the very harmony itself of man with his surroundings, and the completeness of provision for his wants and desires, brought up as a refutation of the validity of the argument for design. It is hard, indeed, if man must be out of harmony with Nature in order to judge anything respecting it, or his relations with it; if he must have experience of chaos before he can predicate anything of order.

But is it true that man has all that he conceives of, or thinks would be useful, and has no "negative evidence of design afforded by the absence of a faculty" to set against the positive evidence afforded by its presence? He notes that he lacks the faculty of flight, sometimes wants it, and in dreams imagines that he has it, yet as thoroughly believes that he was designed not to have it as that he was designed to have the faculties and organs which he possesses. He notes that some animals lack sight, and so, with this negative side of the testimony to the value of vision, he is "apt to infer creative wisdom" both in what he enjoys and in what the lower animal neither needs nor wants. That man does not miss that which he has no conception of, and is by this limitation disqualified from judging rightly of what he can conceive and know, is what the *Westminster Reviewer* comes to, as follows:

"We value the constitution of our world because we live by it, and because we cannot conceive ourselves as living otherwise. Our conceptions of possibility, of law, of regularity, of logic, are all derived from the same source; and as we are constantly compelled to work with these conceptions, as in our increasing endeavors to better our condition and increase our provision we are constantly compelled to guide ourselves by Nature's regulations, we accustom ourselves to look upon these regularities and conceptions as antecedent to all work, even to a Creator's, and to judge of the origin of Nature as we judge of the origin of inventions and utilities ascribable to man. This explains why the argument of design has enjoyed such universal popularity. But that such popularity is no criterion of the argument's worth, and that, indeed, it is no evidence of anything save of an unhappy weakness in man's mental constitution, is abundantly proved by the explanation itself."

Well, the constitution and condition of man being such that he always does infer design in Nature, what stronger presumption could there possibly be of the relevancy of the inference? We do not say of its correctness: that is another thing, and is not the present point. At the last, as has well been said, the whole question resolves itself into one respecting the ultimate veracity of Nature, or of the author of Nature, if there be any.

Passing from these attempts to undermine the foundation of the doctrine—which we judge to be unsuccessful—we turn to the consideration of those aimed at the superstructure. Evidences of design may be relevant, but not cogent. They may, as Mill thought, preponderate, or the wavering balance may incline the other way. There are two lines of argument: one against the sufficiency, the other against the necessity, of the principle of design. Design has been denied on the ground that it squares with only one part of the facts, and fails to explain others; it may be superseded by showing that all the facts are in the way of being explained without it.

The things which the principle of design does not explain are many and serious. Some are in their nature inexplicable, at least are beyond the power and province of science. Others are of matters which scientific students have to consider, and upon which they may form opinions, more or less well-grounded. As to biological science—with which alone we are concerned—it is getting to be generally thought that this principle, as commonly understood, is weighted with much more than it can carry.

This statement will not be thought exaggerated by those most familiar with the facts and the ideas of the age, and accustomed to look them in the face. Design is held to, no doubt, by most, and by a sure instinct; not, however, as always offering an explanation of the facts, but in spite of the failure to do so.

The stumbling-blocks are various, and they lie in every path: we can allude only to one or two as specimens.

Adaptation and utility are the marks of design. What, then, are organs not adapted to use marks of? Functionless organs of some sort are the heritage of almost every species. We have ways of seeming to account for them—and of late one which may really account for them—but they are unaccountable on the principle of design. Some, shutting their eyes to the difficulty, deny that we know them to be functionless, and prefer to believe they must have a use because they exist, and are more or less connected with organs which are correlated to obvious use; but only blindfolded persons care to tread the round of so narrow a circle. Of late some such abortive organs in flowers and fruits are found to have a use, though not the use of their kind. But unwavering believers in design should not trust too much to instances of this sort. There is an old adage that, if anything be kept long enough, a use will be found for it. If the following up of this line, when it comes in our way, should bring us round again to a teleological principle, it will not be one which conforms to the prevalent ideas now attacked

It is commonly said that abortive and useless organs exist for the sake of symmetry, or as parts of a plan. To say this, and stop there, is a fine instance of mere seeming to say something. For, under the principle of design, what is the sense of introducing useless parts into a useful organism, and what shadow of explanation does "symmetry" give? To go further and explain the cause of the symmetry and how abortive organs came to be, is more to the purpose, but it introduces quite another principle than that of design. The difficulty recurs in a somewhat different form when an organ is useful and of exquisite perfection in some species, but functionless in another. An organ, such as an eye, strikes us by its exquisite and, as we say, perfect adaptation and utility in some animal; it is found repeated, still useful but destitute of many of its adaptations, in some animal of lower grade; in some one lower still it is rudimentary and useless. It is asked, If the first was so created for its obvious and actual use, and the second for such use as it has, what was the design of the third? One more case, in which use after all is well subserved, we cite from the article already much quoted from:

"It is well known that certain fishes (*Pleuronecta*) display the singularity of having both eyes on the same side of their head, one eye being placed a little higher than the other. This arrangement has its utility; for the *Pleuronecta*, swimming on their side quite near the bottom of the sea, have little occasion for their eyesight except to observe what is going on above them. But the detail to which we would call notice is, that the original position of the eyes is symmetrical in these fishes, and that it is only at a certain point of their development that the anomaly is manifested, one of the eyes passing to the other side of the head. It is almost inconceivable that an intelligent being should have selected such an arrangement; and that, intending the eyes to be used only on one side of the head, he should have placed them originally on different sides."

Then the waste of being is enormous, far beyond the common apprehension. Seeds, eggs, and other germs, are designed to be plants and animals, but not one of a thousand or of a million achieves its destiny. Those that fall into fitting places and in fitting numbers find beneficent provision, and, if they were to wake to consciousness, might argue design from the adaptation of their surroundings to their well-being. But what of the vast majority that perish? As of the light of the sun, sent forth in all directions, only a minute portion is intercepted by the earth or other planets where some of it may be utilized for present or future life, so of potential organisms, or organisms begun, no larger proportion attain the presumed end of their creation.

"Destruction, therefore, is the rule; life is the exception. We notice chiefly the exception—namely, the lucky prize-winner in the lottery—and take but little thought about the losers, who vanish from our field of observation, and whose number it is often impossible to estimate. But, in this question of design, the losers are important witnesses. If the maxim '*audi alteram partem*' is applicable anywhere, it is applicable here. We must hear both sides, and the testimony of the seed fallen on good ground must be corrected by the testimony of that which falls by the wayside, or on the rocks. When we find, as we have seen above, that the sowing is a scattering at random, and that, for one being provided for and living, ten thousand perish unprovided for, we must allow that the existing order would be accounted as the worst disorder in any human sphere of action."

It is urged, moreover, that all this and much more applies equally to the past stages of our earth and its immensely long and varied succession of former inhabitants, different from, yet intimately connected with, the present. It is not one specific creation that the question has to deal with—as was thought not very many years ago—but a series of creations through countless ages, and of which the beginning is unknown.

These references touch a few out of many points, and merely allude to some of the difficulties which the unheeding pass by, but which, when brought before the mind, are seen to be stupendous.

Somewhat may be justly, or at least plausibly, said in reply to all this from the ordinary standpoint, but probably not to much effect. There were always insuperable difficulties, which, when they seemed to be few, might be regarded as exceptional; but, as they increase in number and variety, they seem to fall into a system. No doubt we may still insist that, "in the present state of our knowledge, the adaptations in Nature afford a large balance of probability in favor of creation by intelligence," as Mill concluded; and probability must needs be the guide of reason through these dark places. Still, the balancing of irreconcilable facts is not a satisfying occupation, nor a wholly hopeful one, while fresh weights are from time to time dropping into the lighter side of the balance. Strong as our convictions are, they may be overborne by evidence. We cannot rival the fabled woman of Ephesus, who, beginning by carrying her calf from the day of its birth, was still able to do so when it became an ox. The burden which our fathers carried comfortably, with some adventitious help, has become too heavy for our shoulders.

Seriously, there must be something wrong in the position, some baleful error mixed with the truth, to which this contradiction of our inmost convictions may be attributed. The error, as we suppose, lies in the combination of the principle of design with the hypothesis of the immutability and isolated creation of species. The latter hypothesis, in its nature unprovable, has, on scientific grounds, become so far improbable that few, even of the anti-Darwinian naturalists, now hold to it; and, whatever may once have been its religious claims, it is at present a hinderance rather than a help to any just and consistent teleology.

By the adoption of the Darwinian hypothesis, or something like it, which we incline to favor, many of the difficulties are obviated, and others diminished. In the comprehensive and far-reaching teleology which may take the place of the former narrow conceptions, organs and even faculties, useless to the individual, find their explanation and reason of being. Either they have done service in the past, or they may do service in the future. They may have been essentially useful in one way in a past species, and, though now functionless, they may be turned to useful account in some very different way hereafter. In botany several cases come to our mind which suggest such interpretation.

Under this view, moreover, waste of life and material in organic Nature ceases to be utterly inexplicable, because it ceases to be objectless. It is seen to be a part of the general "economy of Nature," a phrase which has a real meaning. One good illustration of it is furnished by the pollen of flowers. The seeming waste of this in a pine-forest is enormous. It gives rise to the so-called "showers of sulphur," which every one has heard of. Myriads upon myriads of pollen-grains (each an elaborate organic structure) are wastefully dispersed by the winds to one which reaches a female flower and fertilizes a seed. Contrast this with one of the close-fertilized flowers of a violet, in which there are not many times more grains of pollen produced than there are of seeds to be fertilized; or with an orchis-flower, in which the proportion is not widely different. These latter are certainly the more economical; but there is reason to believe that the former arrangement is not wasteful. The plan in the violet-flower assures the result with the greatest possible saving of material and action; but this result, being close-fertilization or breeding in and in, would, without much doubt, in the course of time, defeat the very object of having seeds at all.[1] So the same plant produces other flowers also, provided with a large surplus of pollen, and endowed (as the others are not) with color, fragrance, and nectar, attractive to certain insects, which are thereby induced to convey this pollen from blossom to blossom, that it may fulfill its office. In such blossoms, and in the great majority of flowers, the fertilization and consequent perpetuity of which are committed to insects, the likelihood that much pollen may be left behind or lost in the transit is sufficient reason for the apparent superfluity. So, too, the greater economy in orchis-flowers is accounted for by the fact that the pollen is packed in coherent masses, all attached to a common stalk, the end of which is expanded into a sort of button, with a glutinous adhesive face (like a bit of sticking-plaster), and this is placed exactly where the

[1] *See* page 346.

head of a moth or butterfly will be pressed against it when it sucks nectar from the flower, and so the pollen will be bodily conveyed from blossom to blossom, with small chance of waste or loss. The floral world is full of such contrivances; and while they exist the doctrine of purpose or final cause is not likely to die out. Now, in the contrasted case, that of pine-trees, the vast superabundance of pollen would be sheer waste if the intention was to fertilize the seeds of the same tree, or if there were any provision for insect-carriage; but with wide-breeding as the end, and the wind which "bloweth where it listeth" as the means, no one is entitled to declare that pine-pollen is in wasteful excess. The cheapness of wind-carriage may be set against the over-production of pollen.

Similar considerations may apply to the mould-fungi and other very low organisms, with spores dispersed through the air in countless myriads, but of which only an infinitesimal portion find opportunity for development. The myriads perish. The exceptional one, falling into a fit medium, is imagined by the *Westminster Reviewer* to argue design from the beneficial provision it finds itself enjoying, in happy ignorance of the perishing or latent multitude. But, in view of the large and important part they play (as the producers of all fermentation and as the omnipresent scavenger-police of Nature), no good ground appears for arguing either wasteful excess or absence of design from the vast disparity between their potential and their actual numbers. The reserve and the active members of the force should both be counted in, ready as they always and everywhere are for service. Considering their ubiquity, persistent vitality, and promptitude of action upon fitting occasion, the suggestion would rather be that, while

> ". . . . thousands at His bidding speed,
> And post o'er land and ocean without rest,
> They also serve [which] only stand and wait."

Finally, Darwinian teleology has the special advantage of accounting for the imperfections and failures as well as for successes. It not only accounts for them, but turns them to practical account. It explains the seeming waste as being part and parcel of a great economical process. Without the competing multitude, no struggle for life; and without this, no natural selection and survival of the fittest, no continuous adaptation to changing surroundings, no diversification and improvement, leading from lower up to higher and nobler forms. So the most puzzling things of all to the old-school teleologists are the *principia* of the Darwinian. In this system the forms and species, in all their variety, are not mere ends in themselves, but the whole a series of means and ends, in the contemplation of which we may obtain higher and more comprehensive, and perhaps worthier, as well as more consistent, views of design in Nature than heretofore. At least, it would appear that in Darwinian evolution we may have a theory that accords with if it does not explain the principal facts, and a teleology that is free from the common objections.

But is it a teleology, or rather—to use the newfangled term—a dysteleology? That depends upon how it is held. Darwinian evolution (whatever may be said of other kinds) is neither theistical nor non-theistical. Its relations to the question of design belong to the natural theologian, or, in the larger sense, to the philosopher. So long as the world lasts it will probably be open to any one to hold consistently, in the last resort, either of the two hypotheses, that of a divine mind, or that of no divine mind. There is no way that we know of by which the alternative may be excluded. Viewed philosophically, the question only is, Which is the better supported hypothesis of the two?

We have only to say that the Darwinian system, as we understand it, coincides well with the theistic view of Nature. It not only acknowledges purpose (in the *Contemporary Reviewer's* sense),[1] but builds upon it; and if purpose in this sense does not of itself imply design, it is certainly compatible with it, and suggestive of it. Difficult as it may be to conceive and impossible to demonstrate design in a whole of which the series of parts appear to be contingent, the alternative may be yet more difficult and less satisfactory. If all Nature is of a piece—as modern physical philosophy insists—then it seems clear that design must in some way, and in some sense, pervade the system, or be wholly absent from it. Of the alternatives, the predication of design—special, general, or universal, as the case may be—is most natural to the mind; while the exclusion of it throughout, because some utilities may happen, many adaptations may be contingent results, and no organic mal-

[1] *See* pp. 358, 359.

adaptations could continue, runs counter to such analogies as we have to guide us, and leads to a conclusion which few men ever rested in. It need not much trouble us that we are incapable of drawing clear lines of demarkation between mere utilities, contingent adaptations, and designed contrivances in Nature; for we are in much the same condition as respects human affairs and those of lower animals. What results are comprehended in a plan, and what are incidental, is often more than we can readily determine in matters open to observation. And in plans executed mediately or indirectly, and for ends comprehensive and far-reaching, many purposed steps must appear to us incidental or meaningless. But the higher the intelligence, the more fully will the incidents enter into the plan, and the more universal and interconnected may the ends be. Trite as the remark is, it would seem still needful to insist that the failure of a finite being to compass the designs of an infinite mind should not invalidate its conclusions respecting proximate ends which he can understand. It is just as in physical science, where, as our knowledge and grasp increase, and happy discoveries are made, wider generalizations are formed, which commonly comprehend, rather than destroy, the earlier and partial ones. So, too, the "sterility" of the old doctrine of final causes in science, and the presumptuous uses made of them, when it was supposed that every adapted arrangement or structure existed for this or that direct and special end, and for no other, can hardly be pressed to the conclusion that there are no final causes, i. e., ultimate reasons of things.[1] Design in Nature is distinguished from that in human affairs—as it fittingly should be—by all comprehensiveness and system. Its theological synonym is Providence. Its application in particular is surrounded by similar insoluble difficulties; nevertheless, both are bound up with theism.

Probably few at the present day will maintain that Darwinian evolution is incompatible with the principle of design; but some insist that the theory can dispense with, and in fact supersedes, this principle.

The *Westminster Reviewer* cleverly expounds how it does so. The exposition is too long to quote, and an abstract is unnecessary, for the argument adverse to design is, as usual, a mere summation or illustration of the facts and assumptions of the hypothesis itself, by us freely admitted. Simplest forms began; variations occurred among them; under the competition consequent upon the arithmetical or geometrical progression in numbers, only the fittest for the conditions survive and propagate, vary further, and are similarly selected; and so on.

"Progress having once begun by the establishment of species, the laws of atavism and variability will suffice to tell the remainder of the story. The colonies gifted with the faculty of forming others in their likeness will soon by their increase become sole masters of the field; but the common enemy being thus destroyed, the struggle for life will be renewed among

[1] "No single and limited good can be assigned by us as the final cause of any contrivance in Nature. The real final cause . . . . is the sum of all the uses to which it is ever to be put. Any use to which a contrivance of Nature is put, we may be sure, is a part of its final cause."—(G. F. Wright, in *The New-Englander*, October, 1871.)

the conquerors. The saying that 'a house divided against itself cannot stand,' receives in Nature its flattest contradiction. Civil war is here the very instrument of progress; it brings about the survival of the fittest. Original differences in the cell-colonies, however slight, will bring about differences of life and action; the latter, continued through successive generations, will widen the original differences of structure; innumerable species will thus spring up, branching forth in every direction from the original stock; and the competition of these species among each other for the ground they occupy, or the food they seek, will bring out and develop the powers of the rivals. One chief cause of superiority will lie in the division of labor instituted by each colony; or, in other words, in the localization of the colony's functions. In the primitive associations (as in the lowest organisms existing now), each cell performed much the same work as its neighbor, and the functions necessary to the existence of the whole (alimentation, digestion, respiration, etc.) were exercised by every colonist in his own behalf. Social life, however, acting upon the cells as it acts upon the members of a human family, soon created differences among them—differences ever deepened by continuance, and which, by narrowing the limits of each colonist's activity, and increasing his dependence on the rest, rendered him fitter for his special task. Each function was thus gradually monopolized; but it came to be the appanage of a single group of cells, or *organ;* and so excellent did this arrangement prove, so greatly were the powers of each commonwealth enhanced by the division of its labor, that the more organs a colony possessed, the more likely it was to succeed in its struggle for life. . . . We shall go no further, for the reader will easily fill out the remainder of the picture for himself. Man is but an immense colony of cells, in which the division of labor, together with the centralization of the nervous system, has reached its highest limit. It is chiefly to this that his superiority is due; a superiority so great, as regards certain functions of the brain, that he may be excused for having denied his humbler relatives, and dreamed that, standing alone in the centre of the universe, sun, moon, and stars, were made for him."

Let us learn from the same writer how both eyes of the flounder get, quite unintentionally, on the same side of the head. The writer makes much of this case (*see* p. 372), and we are not disposed to pass it by:

"A similar application may be made to the *Pleuronecta*. Presumably, these fishes had adopted their peculiar mode of swimming long before the position of their eyes became adapted to it. A spontaneous variation occurred, consisting in the passage of one eye to the opposite side of the head; and this variation afforded its possessors such increased facilities of sight that in the course of time the exception became the rule. But the remarkable point is, that the law of heredity not only preserved the variation itself, but the date of its occurrence; and that, although for thousands of years the adult *Pleuronecta* have had both eyes on the same side, the young still continue during their earlier development to exhibit the contrary arrangement, just as if the variation still occurred spontaneously."

Here a wonderful and one would say unaccountable transference takes place in a short time. As Steenstrup showed, one eye actually passes through the head while the young fish is growing. We ask how this comes about; and we are told, truly enough, that it takes place in each generation because it did so in the parents and in the whole line of ancestors. Why offspring should be like parent is more than any one can explain; but so it is, in a manner so nearly fixed and settled that we can count on it; yet not from any absolute necessity that we know of, and, indeed, with sufficiently striking difference now and then to demonstrate that it might have been otherwise, or is so in a notable degree. This transference of one eye through the head, from the side where it would be nearly useless to that in which it may help the other, bears all the marks of purpose, and so carries the implication of design. The case is adduced as part of the evidence that Darwinian evolution supersedes design. But how? Not certainly in the way this goes on from generation to generation; therefore, doubtless in the way it began. So we look for the explanation of how it came about at the first unintentionally or accidentally; how, under known or supposed conditions, it must have happened, or at least was likely to happen. And we read, "A spontaneous variation occurred, consisting in the passage of one eye to the opposite side of the head." That is all; and we suppose there is nothing more to be said. In short, this surprising thing was undesigned because it took place, and has taken place ever since! The writer presumes, moreover (but this is an *obiter dictum*), that the peculiarity originated long after flounders had fixed the habit of swimming on one side (and in this particular case it is rather difficult to see how the two may have gone on *pari passu*), and so he cuts away all obvious occasion for the alteration through the summation of slight variations in one direction, each bringing some advantage.

This is a strongly-marked case; but its features, although unusually prominent, are like those of the general run of the considerations by which evolution is supposed to exclude design. Those of the penultimate citation and its context are all of the same stamp. The differences which begin as variations are said to be spontaneous—a metaphorical word of wide meanings—are inferred to be casual (whereas we only know them to be occult), or to be originated by surrounding agencies (which is not in a just sense true); they are legitimately inferred to be led on by natural selection, wholly new structures or organs appear, no one can say how, certainly no one can show that they are necessary outcomes of what preceded; and these two are through natural selection kept in harmony with the surroundings, adapted to different ones, diversified, and perfected; purposes are all along subserved through exquisite adaptations; and yet the whole is thought to be undesigned, not because of any assigned reason why this or that must have been thus or so, but simply because they all occurred in Nature! The Darwinian theory implies that the birth and development of a species are as natural as those of an individual, are facts of the same kind in a higher order. The alleged proof of the absence of design from it amounts to a simple reiteration of the statement, with particulars. Now, the marks of contrivance in the structure of animals used not to be questioned because of their coming in the way of birth and development. It is curious that a further extension of this birth and development should be held to disprove them. It appears to us that all this is begging the question against design in Nature, instead of proving that it may be dispensed with.

Two things have helped on this confusion. One is the notion of the direct and independent creation of species, with only an ideal connection between them, to question which was thought to question the principle of design. The other is a wrong idea of the nature and province of natural selection. In former papers we have over and over explained the

Darwinian doctrine in this respect. It may be briefly illustrated thus: Natural selection is not the wind which propels the vessel, but the rudder which, by friction, now on this side and now on that, shapes the course. The rudder acts while the vessel is in motion, effects nothing when it is at rest. Variation answers to the wind: "Thou hearest the sound thereof, but canst not tell whence it cometh and whither it goeth." Its course is controlled by natural selection, the action of which, at any given moment, is seemingly small or insensible; but the ultimate results are great. This proceeds mainly through outward influences. But we are more and more convinced that variation, and therefore the ground of adaptation, is not a product of, but a response to, the action of the environment. Variations, in other words, the differences between individual plants and animals, however originated, are evidently not from without but from within—not physical but physiological.

We cannot here assign particularly the reasons for this opinion. But we notice that the way in which varieties make their appearance strongly suggests it. The variations of plants which spring up in a seed-bed, for instance, seem to be in no assignable relation to the external conditions. They arise, as we say, spontaneously, and either with decided characters from the first, or with obvious tendencies in one or few directions. The occult power, whatever it be, does not seem in any given case to act vaguely, producing all sorts of variations from a common centre, to be reduced by the struggle for life to fewness and the appearance of order; there are, rather, orderly indications from the first. The variations of which we speak, as originating in no obvious causal relation to the external conditions, do not include dwarfed or starved, and gigantesque or luxuriant forms, and those drawn up or expanded on the one hand, or contracted and hardened on the other, by the direct difference in the supply of food and moisture, light and heat. Here the action of the environment is both obvious and direct. But such cases do not count for much in evolution.

Moreover, while we see how the mere struggle and interplay among occurring forms may improve them and lead them on, we cannot well imagine how the adaptations which arrest our attention are thereby secured. Our difficulty, let it be understood, is not about the natural origination of organs. To the triumphant outcry, "How can an organ, such as an eye, be formed under Nature?" we would respond with a parallel question, How can a complex and elaborate organ, such as a nettle-sting, be formed under Nature? But it is so formed. In the same species some individuals have these exquisitely-constructed organs and some have not. And so of other glands, the structure and adaptation of which, when looked into, appear to be as wonderful as anything in Nature. The impossibility lies in conceiving how the obvious purpose was effectuated under natural selection alone. This, under our view, any amount of gradation in a series of forms goes a small way in explaining. The transit of a young flounder's eye across the head is a capital instance of a wonderful thing done under Nature, and done unaccountably.

But simpler correlations are involved in similar difficulty. The superabundance of the pollen of pine-trees above referred to, and in oak-trees, is correlated with chance fertilization under the winds. In the analogous instance of willows a diminished amount of pollen is correlated with direct transportation by insects. Even in so simple a case as this it is not easy to see how this difference in the conveyance would reduce the quantity of pollen produced. It is, we know, in the very alphabet of Darwinism that if a male willow-tree should produce a smaller amount of pollen, and if this pollen communicated to the offspring of the female flowers it fertilized a similar tendency (as it might), this male progeny would secure whatever advantage might come from the saving of a certain amount of work and material; but why should it begin to produce less pollen? But this is as nothing compared with the arrangements in orchid-flowers, where new and peculiar structures are introduced—structures which, once originated and then set into variation, may thereupon be selected, and thereby led on to improvement and diversification. But the origination, and even the variation, still remains unexplained either by the action of insects or by any of the processes which collectively are personified by the term natural selection. We really believe that these exquisite adaptations have come to pass in the course of Nature, and under natural selection, but not that natural selection alone explains or in a just sense originates them. Or rather, if this term is to stand for sufficient cause and rational explanation, it must denote or include that inscrutable something which produces—as well as that which results in the survival of—"the fittest."

We have been considering this class of questions only as a naturalist might who sought for the proper or reasonable interpretation of the problem before him, unmingled with considerations from any other source. Weightier arguments in the last resort, drawn from the intellectual and moral constitution of man, lie on a higher plane, to which it was unnecessary for our particular purpose to rise, however indispensable this be to a full presentation of the evidence of mind in Nature. To us the evidence, judged as impartially as we are capable of judging, appears convincing. But, whatever view one unconvinced may take, it cannot remain doubtful what position a theist ought to occupy. If he cannot recognize design in Nature because of evolution, he may be ranked with those of whom it was said, "Except ye see signs and wonders ye will not believe." How strange that a convinced theist should be so prone to associate design only with miracle!

All turns, however, upon what is meant by this Nature, to which it appears more and more probable that the being and becoming—no less than the well-being and succession—of species and genera, as well as of individuals, are committed. To us it means "the world of force and movement in time and space," as Aristotle defined it—the system and totality of things in the visible universe.

What is generally called Nature Prof. Tyndall names matter—a peculiar nomenclature, requiring new definitions (as he avers), inviting misunderstanding, and leaving the questions we are concerned with just where they were. For it is still to ask: whence this rich endowment of matter? Whence comes that of which all we see and know is the outcome? That to which potency may in the last resort be ascribed, Prof. Tyndall, suspending further judgment, calls mystery—using the word in one of its senses, namely, something hidden from us which we are not to seek to know. But there are also mysteries proper to be inquired into and to be reasoned about; and, although it may not be given unto us to *know* the mystery of causation, there can hardly be a more legitimate subject of philosophical inquiry. Most scientific men have thought themselves intellectually authorized to have an opinion about it. "For, by the primitive and very ancient men, it has been handed down in the form of myths, and thus left to later generations, that the *Divine* it is which holds together all Nature;" and this tradition, of which Aristotle, both naturalist and philosopher, thus nobly speaks[1]—continued through succeeding ages, and illuminated by the Light which has come into the world—may still express the worthiest thoughts of the modern scientific investigator and reasoner.

[1] *Παραδέδοται δὲ ὑπὸ τῶν ἀρχαίων καὶ παμπαλαίων ἐν μύθου σχήματι καταλελειμένα τοῖς ὕστερον, ὅτι περιέχει* ΤΟ ΘΕΙΟΝ *τὴν ὅλην φύσιν.*—*Arist. Metaphys.*, xi. 8, 19.

INDEX.

THE END.

Printed in the United States
By Bookmasters